OUR BRAINS, OUR SELVES

OUR BRAINS, OUR SELVES

What A Neurologist's Patients Taught Him About the Brain

MASUD HUSAIN

CANONGATE

First published in Great Britain in 2025
by Canongate Books Ltd, 14 High Street, Edinburgh EH1 1TE

canongate.co.uk

4

British Library Cataloguing-in-Publication Data
A catalogue record for this book is available on
request from the British Library

ISBN 978 1 80530 105 9

Typeset in Bembo Std by Palimpsest Book Production Ltd,
Falkirk, Stirlingshire

Printed and bound by CPI Group (UK) Ltd, Croydon CR0 4YY

The manufacturer's authorised representative in the EU for product safety is BGC
Sustainability & Compliance, 7 avenue du Général Leclerc, Paris 75014
(gpsr@baldwinglobalconsulting.com)

For my family and all those who made it possible for me to get here

Contents

Preface

What makes us who we are? Most of us might say that it is our background that creates our identities: our families, where we've lived, how we were brought up and educated, the people who have influenced us, the jobs we've held. But there is something far more fundamental that makes us who we are, and which transcends social and cultural experiences. This is our brain. Our brains create us. No matter where you are from, where you live or have lived, the language you speak, the colour of your skin, it is our brains that give us our identities.

In the past, some disagreed, arguing instead as Descartes did that our personal identity – our '*self*' – is separate from the brain. Most modern views, however, consider the brain to be the basis for all the experiences we have of our selves. Using new scanning techniques, some neuroscientists have even attempted to define the brain region where 'the self' might reside. However, as the philosopher Daniel Dennett drily put it: 'It is a category mistake to start looking around for the self in the brain.' We're not going to find it because our self is made up by the entirety of our brain functions.

Indeed, many thinkers have proposed that the self is simply

1

an illusion or even a fictional narrative created by our brains. It doesn't exist, so therefore it cannot be localised to any particular brain region. Others have argued that there are multiple selves: the processes we attribute to a unitary self actually reflect the workings of a distributed mechanism. The self is simply an emergent property of our entire brain. To paraphrase the artificial intelligence pioneer Marvin Minsky, the self is simply the product of the many different cognitive processes that make up the 'society of our minds'.

That this is the case is made starkly real when we lose even one aspect of our cognitive abilities, one aspect of that 'society': such as our memory or motivation, perception or language, the ability to pay attention, make good decisions, empathise with other people, plan or think ahead. Then it may become apparent that we are not the same person we used to be, that we have lost a piece of our selves – our *personal identity*.

People who develop a brain disorder can experience a profound change not only in their personal identity but also in their *social identity*. They can be transformed into almost unrecognisable individuals, effectively becoming outsiders even to close friends and family, finding it extremely challenging to remain in their social networks.

This is a book about encounters I've had as a practising neurologist and neuroscientist with such individuals. It's a book about what these patients tell us about how human brains work and how they create our personal and social identities – our *selves*. In each chapter we meet someone who has become different because of their neurological disorder. We'll find out what happened to them, how they were affected and why they changed – the neuroscience behind their transformation.

Most importantly, we'll discover what that change tells us about how our brains normally function to make us who we are. We'll see how our constellations of connections with other people depend on the operations of many different cognitive functions, which contribute to creating our personal as well as our social identities, ultimately allowing us to 'belong' within our social networks. Finally, we'll discover how modern neuro-science is offering hope to people with brain diseases.

Introduction

Dawn had crept serenely over the city. The shadows draped over the avenues were slowly receding to usher in a beautiful, bright morning. It was June, and the few early risers on their way to set up market stalls were basking in the quiet, pale radiance of the new day, some small comfort when the enemy was only fifty miles away. Many who could afford to had already fled the metropolis, but most of the inhabitants clung to the belief that the defensive line would hold, as it had for nearly four years. There was still hope.

On Boulevard Haussmann, a few cars headed eastwards but otherwise the street was quiet, with most of its inhabitants still to awaken. The occupant of the second-floor apartment at number 102, though, had been up for some time – all night, in fact. The shutters of his windows were tightly drawn, as they had been for months. His green bedside lamp was the only source of light in the gloomy bedroom. Crammed full of dark furniture with books piled high on the desk, and the heady vapours of stramonium for his asthma engulfing the chamber with pungent fumes, the room held an air of oppressive confinement. Its cork-lined walls, especially installed to isolate the occupant from the sounds of the street and the rest of the building, added to the sense of claustrophobia most of his visitors must have felt.[1]

Sitting up in bed in his ornate Japanese housecoat and propped up on two large pillows, he would at this time of the day normally be working feverishly on his manuscript, which he had been writing diligently by hand in black leather notebooks over the last twelve years. But this morning was different. He had been gripped by an overwhelming fear. One side of his face, he was sure, was drooping. When he had spoken to his housekeeper, Céleste, on the previous evening he was convinced that his utterances had been slurred, his speech somehow garbled. He must be on the verge of suffering a major stroke, he concluded, just as both his mother and father had been afflicted. There could be no other explanation. It was in the family blood. And hadn't his beloved mother, Jeanne, been left with a terrible infirmity? Her stroke had robbed her of language: she had become aphasic, unable to talk to her precious sons.

So it was that in the summer of 1918, as the Germans were launching their final offensives of the First World War with the aim of reaching Paris, that the great novelist Marcel Proust sat in his blue satin sheets contemplating with dread the possibility of a brain disorder, one that would deprive him of his most cherished ability: to communicate. Now in his late forties, he was highly familiar with aphasia. Not only had his mother suffered from it but, before his own stroke, his father Dr Adrien Proust had written an entire book on the topic.

The younger Marcel had also made the acquaintance of many of the most accomplished neurologists in the city.[2] At that time, Paris was considered to be the leading centre for brain disorders in the world, with several of its pioneering experts having made landmark contributions to the subject. These included developing an understanding of disorders of language after stroke, which can impair not only the ability to speak, but also to read and write. Without these faculties, where would Proust be?

Such was the dread of his impending aphasia that morning in June 1918 that he made an appointment to see the celebrated neurologist, Joseph Babinski, whose consulting chambers were located just ten minutes away at number 170 on the same boulevard. As Proust recalled the encounter, Babinski had no knowledge of him. 'Do you have a job?' Babinski had apparently asked.[3]

Proust's objective that day was to get Babinski to perform a trepanation: to make a hole in his skull. So great was his fear that he was convinced that such a radical course of action was necessary to prevent a stroke from progressing. Babinski, ever the professional, examined Proust and reassured him that there was no evidence that he was suffering a stroke and gently declined to perform the surgical procedure.[4] Goodness knows what might have happened to Proust's great novel if he had. Marcel Proust never did experience a stroke, although the anxiety of being struck down by one continued to plague him inter- mittently for the rest of his short life. Even when, a few years later, he was dying from pneumonia, it was Babinski who was called for.

Proust's concerns about suffering from a condition that affects the brain are not unique. Although any one of us can develop an illness that affects our bodies, what many of us fear the most is a disorder that affects our brains. Why? Because neurological conditions can make people become so very different.[5] Some may not be able to communicate, as Proust had feared. Others might lose their memory or suffer from distorted perceptions or hallucinations. Some may become socially inappropriate, lacking in empathy, or be rude and aggressive. Others might become very impulsive or disinhibited, gambling away large amounts of money or developing new addictions. Some might suffer from pathological apathy, becoming withdrawn and lacking motivation to interact with other people.

Understandably, alterations in behaviour or personality like these can be extremely upsetting and frightening to people who develop them and to their families. But they also reveal a lot about you and me. By observing what happens when a particular brain function is lost, we can learn a great deal about our normal *selves*, how cognitive functions contribute to create who we are (our *personal identities*), as well as how they shape our *social identities* – the part of our selves that derives from our relationships to others.

For someone like Marcel Proust, loss of language would have been calamitous. Not only would he lose the ability to write but, perhaps just as importantly, he would no longer have the same presence within his social circle. The social identity that he had worked so hard to craft would effectively dissolve. Proust had spent years cultivating relationships with some of the loftiest members of French society. He had an inordinate preoccupation with his relationship to people of influence. For a man who was gay and from a Jewish background (on his mother's side), he had deftly managed to navigate the complexities of Parisian prejudice and snobbery with enormous success.

Through observation and emulation, he had become an insider in a world where few would have thought he belonged or would have any sway. Indeed, some commentators have concluded that Proust was a highly effective manipulator, a man who was unwilling to relinquish his own influence over others in his orbit, even when he spent days on end writing in his sombre bedroom.[6] Without language, though, the circles that he had worked so hard to be part of would no longer be accessible. He wouldn't 'belong'.

~ o ~

Have you ever considered who you truly are and whether you belong? Have you caught yourself wondering how you might get to fit into a particular group? I have. I suspect that most of us have. The term 'to belong' has an interesting origin. The suffix 'long' stems from an Old English word that means to yearn for, while the prefix 'be' denotes closeness. To long to be close to others, to fit in and be part of a group is a common human desire.

Indeed, some psychologists have proposed that humans have a fundamental need to belong.[7] It is a primary motivation, they argue, for much of what we do, a universal driver of our cultures which makes relationships and bonds between people such a prominent part of human existence. Belonging might also be important because it gives groups an evolutionary advantage. By sharing food and knowledge, by hunting or working together, or even caring for children as a community, human beings collectively can solve problems more effectively. Being part of a social group matters for our success and well-being.[8]

In the 1950s, psychologists Manfred Kuhn and Thomas McPartland devised a very direct test in an attempt to discover how people define themselves. They simply asked their participants to write down up to twenty statements that describe who they are by completing, each time, the following sentence: 'I am . . .'[9] The results revealed that young American students tended to base their descriptions of themselves more on the groups they belonged to (e.g. 'I am a Catholic' or 'I am African American') than on their individual traits (e.g. 'I am happy' or 'I am bored'). Decades later, there was a shift towards responding more with descriptions that described character and personality traits rather than group affiliations.[10] Nevertheless, the social groups to which people belonged remained a key component of how they viewed their identity. These groups help to define who we are.

Right from childhood, the urge to be part of the 'in crowd' seems compelling, somehow comforting for us. Even outside of our circle of friends, most of us consider that we are part of something more than just ourselves: we belong to a family, a partnership, an organisation we work for, a region where we grew up, perhaps even a 'tribe' that we are members of, a language we speak, a sports team we support, a religion we follow, a nation we are part of, or even a group of nations – a people with a common cultural heritage – that we are associated with.

Most of us feel we belong to several of these collections, and our constellation of attachments helps to define who we are. These constellations also allow other people to categorise us, to have a compact description of our presumed backgrounds and to lay down assumptions of how we are likely to behave. They help to brand us with a cultural and social type – to give us some form of 'surface' identity. This might not always be a true characterisation of who we are. Not everyone fits a stereotype. Nevertheless, these classifications have, for centuries, dominated how we conventionally start to cast others, as well as ourselves.

Encountering someone with one or more of our own group identities, such as meeting a fellow countryman at a party in another part of the world, brings with it an automatic sense of bond, a common ground, even though we might have no other shared past experiences. Somehow, we identify with them even though we don't know them. We presume that we share with them some aspects of our own identity.

Belonging also means that we are aware of people who are outsiders: those who are not from our tribes or collectives, individuals who don't fit in any of our groups. They can be new to our world, from another region of the country with a different accent, or have a different religion, or perhaps be entirely foreign, migrants who might not even speak the way we do. These people are, by definition, different. A common response to them is to consider them with caution, perhaps even suspicion, simply because they are outsiders. Often, we exclude them. We might discover what it is like to be treated in this way if we spend time away from where we grew up, in a different part of the country or, perhaps more strikingly, if we go to live abroad.

My own personal early experiences have brought a long-standing familiarity with the concept of not belonging, of being an outsider. Many years ago, I was attending my first international scientific conference in the United States. I had to give a short presentation in front of a very large audience. I recall the excitement of anticipation. I had some new research findings on cognitive function in people who had suffered a stroke, and there were hundreds of neuroscientists there, all eagerly waiting to hear me – or so I thought. As I got to the podium and looked around, I could see how packed the room was. It was exhilarating but also extremely daunting. The delivery had to be just right. I had to remember to speak slowly and ensure that I paused at the right places, I told myself.

The chairman introduced me, and I was about to start speaking when I realised to my dismay that most people in the audience were still chatting to each other, engrossed in their conversations. They weren't stopping to look at me and they didn't seem to be at all interested in what I was going to talk about. I had got this all wrong. It was going to be a big disappointment.

I started to speak, thinking I couldn't wait to get this over with. Suddenly the room went silent, people were turning their gaze towards me and, to my delight, seemed to be engrossed by the presentation. By the time I finished and had received a very generous round of applause, I felt elated. As I sat down, a colleague of mine made me feel even better by saying it had been a fantastic talk. I was beaming and told him that I was so pleased the data had been received well.

'You do know it wasn't just that, though,' he said.

'What do you mean?' I asked.

'Well, the reason they were so interested is that they just couldn't believe that posh British accent coming out of a brown guy like you,' he said with a chuckle. 'You're incongruous: the voice doesn't fit the appearance! They were intrigued because *you* don't fit. They thought you were going to be Mexican when they saw you walk up to the podium.'

It was true that almost all of the audience had been white, but I was used to that. I was nevertheless shocked by his obser- vation about my accent. On reflection, I shouldn't have been. I had lots of experience of people considering me to be different, but this comment on the way I spoke was particularly ironic.

Born in East Pakistan (now Bangladesh), I had come to the UK as a child in 1968, some twenty-one years after the British Empire had precipitously left the Indian subcontinent to be run by its own population. Britain had been on its knees follow- ing the Second World War and, pushed by its American allies to whom an empire seemed anti-democratic, it rushed into a hasty, chaotic departure. The post-colonial world had seen many convulsions after the bloody Partition of India. One of the conse- quences, many years later, was that people from all parts of the Empire as it had been – the 'Commonwealth' as it came to be branded – migrated to its very heart.

I grew up in inner-city London and Birmingham at a time when new arrivals from other parts of the world were not warmly embraced. We were considered by the locals to be people who didn't belong. At that time, they sneered at the way I spoke, or called me names as I walked down the street. I didn't know the meaning of some of those words they used, so I'd look them up in a dictionary, frequently finding that the definitions escaped my understanding. But other terms seemed yet more perplexing. With a child's logic, it seemed a truism to be called a 'Paki' when you had in fact been born in Pakistan. Soon it became clear, though, that this was not simply a helpful means of reacquainting me with my origins, but rather a term of hate, capturing in two syllables a vicious bigotry.

Much later, when I got to university, I decided I wanted to be a neurologist. It seemed the only specialty to go into when I was a medical student. At Oxford, we were taught the bewildering complexities of the nervous system. I learned about its beautiful anatomy through dissecting human brains, looking down microscopes and studying books that mapped its complex connections. I found out about its intricate physiology through experiments conducted on nerves and muscles, and by reading scientific papers on electrical recordings from different brain regions. I was taught its elaborate biochemistry and pharmacology, the drugs that affect its functions and even the diseases that could destroy it. But I also learned how, despite all this immense knowledge, acquired over centuries, we understand so little about the nervous system.

The brain in particular was perplexing, as were the disorders that affected it. It is one thing to know how neurons operate, their structure and function, but quite another to understand why humans behave the way they do or how brain diseases affect people and their personalities. There was, however, one little problem that stood in the way of my becoming a brain specialist. In those days, there were only two hundred neurologists covering the entire United Kingdom. British Neurology was an elite club. As a friend of mine put it bluntly: 'You're brown. There isn't a brown neurologist. You're an outsider and you're not going to make it. Try rheumatology; they're far less selective.'

By telling me I couldn't become a neurologist, my friend was reminding me who I really was. Even if I had made it to Oxford as a medical student, there were limits to what British society, even British professional society, was willing to tolerate from an outsider like me. Thankfully, I ignored his advice and persisted.

Although it wasn't always easy, I met many in that elite British Neurology club who were sufficiently open-minded to give me an opportunity, even encourage me. Eventually, I did become a neurologist. Indeed, some might say that I'm now a quintessential insider; as an Oxford professor I am unavoidably part of the academic and medical establishment.

It's true that I no longer feel I don't belong. Through years of observation and emulation, I have fitted very neatly into the customs and patterns of behaviour expected of me. I now speak without a Pakistani (or Mexican) accent – which is what had surprised the audience at my first international conference. But becoming a neurologist proved to be very different from what I had expected. The spectacular, rare conditions had, of course, caught my eye. However, what also became very apparent is that even common neurological conditions can reveal an enormous amount about how our sense of self is normally created by our brains. As we'll see in this book, brain disorders can transform both the personal and social identities of individuals. The way in which people behave can be radically altered, sometimes shockingly so. Indeed, brain diseases can have such a huge impact on daily lives that patients are sometimes turned into people who no longer belong within their social groups.

For decades, social psychologists and anthropologists have studied the demarcations between who is in a group and who is outside of it. The investigation of social identity includes attempts to understand how a person identifies with a particular group's values through shared experiences, and how they establish and maintain boundaries between *them* (the 'out-group') and *us* (the 'in-group').[11] Kinship, language, place of birth, religion, ethnicity, nationality, political views, sexual orientation and social class are only some of the attributes that seem crucial to how the lines are drawn between members of a group and those who don't belong.

What has become apparent, though, is that social identity is established not only by perceiving similarities with another person, but also by determining how we differ from someone else – by contrasting ourselves to *them* (those who we perceive as the 'out-group'). Without 'others' who are different from us, we can't claim to be part of a group with a particular identity. It is only in relation to other people that it is possible to define where we sit, within or outside their group, and how we categorise ourselves. People who do not conform to our in-group expectations are considered outsiders.

This also may be the case for people who once were insiders but now have developed brain disorders which make them behave differently. They too can become part of the 'others', because one of the cognitive functions that helped them to belong (such as attention, perception, memory, motivation, decision-making, empathy) has become dysfunctional. In this way, these individuals can reveal how fundamental cognitive processes are essential to creating our social identities, and the way our sense of self is defined partly by our relationships to other people.

In fact, even for some healthy people who choose to leave 'normal' societal life in order to lead a hermetic existence, being alone can paradoxically lead to a loss of a sense of self. The writer Neil Ansell isolated himself for five years in a cottage in deepest Wales[12] and found that 'alone, there was no need for identity, or self-definition'. Instead, he reflected that his sense of self had come through his interactions with other people. Without this, he concluded, he was losing his identity.

The psychological study of social identity has also focused on how membership of a social group can shift under other circumstances. For example, by marrying a person from a different background, or by moving to a different country and learning the way people behave or speak there, it is possible to move from being an outsider to gaining insider status. On the other hand, individuals can be ostracised from their social groups. Rejection – the loss of insider status – is a practice that stretches back millennia.[13]

The Greeks used it to remove people. The term 'ostracism' comes from them: voters wrote the names of individuals who they wanted to banish on small shards of pottery called *ostrakon*. Pathan hill tribes in Pakistan still use exile as a form of punishment to avert feuding. Among the Amish, the procedure of *meidung* involves individuals being ignored in order to discipline them. Less formally, many societies – even children's playground societies – adopt the practice of shunning those who no longer conform to societal norms, making them feel that they are outsiders. Social identities within any culture can clearly change. They can alter under new circumstances and situations.

But identities and a sense of belonging are not only set by our social and cultural backgrounds. Our brains, and more specifically how they function in everyday life, are crucial. They construct our social identities, determining how we behave in different contexts and with different people. Brains can also allow us to change identities. An outsider, such as an immigrant like me, can eventually become an insider but only if they develop skills that make it possible for them to fit the criteria for group membership. And how does someone acquire those skills? Their brains have to learn the characteristics of a particular group and be able to adapt to these. Perhaps it might be the way people speak – their accent – or their sense of humour, the food or music they prefer, the way they choose to interpret news, or their support for a particular political party or sports team. Regardless of the attribute, some people's brains are capable, through their cognitive abilities, of adapting to new challenges and circumstances to allow them to gain insider status – to 'fit' into a group – and, in so doing, their sense of self also changes.

Cognitive functions can also be lost, sometimes dramatically, with the onset of brain diseases. The people we'll meet in this book, for example, will show us that our insider status and social identity are not fixed either. These individuals were transformed by alterations in their brains brought about by conditions such as stroke, head injury or different types of dementia. When they lost a specific cognitive function, they risked being shunned by people they knew well, of moving from being an insider in a social network that they had belonged to for many years, to becoming someone one who no longer fits.

Encountering patients who have been transformed in this way by different types of neurological disorder raises many questions. A recurring one for me has been: what does it mean to belong? It is extraordinary that someone who has grown up 'belonging' to a group can lose this, sometimes overnight. It can happen to any of us. Our identities can change to leave an altered self, and the way our identity comes across to other people determines whether they consider us 'to belong' or not.

Without certain fundamental brain functions, the personal and social identities we have developed over many years – and projected to others – count for very little. Those identities – the 'self' – that we created and the affiliations we nurtured with different groups over time relied on our brains. The strength of those connections, the quality of the 'in-group' bonds, depends on how our brains function in different situations, in chance conversations or formal exchanges, when confronted with new problems to solve, or when having fun with a group of people. With the onset of a neurological condition that takes away a fundamental cognitive function, we can lose all that our brains have worked so hard to achieve over many years: to forge the connections that gained us access to – and maintained – our group memberships.

It turns out that there are many ways in which this can happen. Neurological disorders, affecting different parts of the brain and caused by a myriad of pathologies (strokes, neurodegeneration, head injury, tumours and so on),[5] can make people behave in strikingly different ways, so different that we no longer consider them to be part of *us*. Patients with these conditions can teach us not only what we take for granted about our selves but also about how human brains build a sense of identity and of belonging over time. That seemingly unified sense of 'self', our feeling of who we are and what we belong to, turns out to be far more complex than most of us might ever imagine.[14] There are, in effect, many different parts to our identities because, as neurological conditions reveal, many different brain functions contribute to construct our *selves*.

Across seven chapters, we'll reflect on the neuroscience of identity and belonging, learning about seven different people whose lives were transformed in very different ways by their brain disorders. Each loses a key cognitive faculty that leads them to risk becoming an outsider. We'll consider what these people reveal about how our brains construct our identities and how cognitive functions can determine whether we belong within a group or not. Although shifts in group membership are clearly possible, regaining insider status for patients with neurological conditions can be a formidable challenge. Nonetheless, as we shall see, the advent of new treatments is paving the way for better outcomes and bringing hope to patients and families around the world.

To revert to our original question: what makes us who we are? Of course, our backgrounds – families, friends, upbringing, jobs, education, homes – are important in creating our identities. These all influence who we are and who we become. But the people we'll meet in this book will reveal how our personal and social identity – our *self* – is fundamentally made possible by the functions of many different cognitive processes in our brains. Together these cognitive functions create us. They are the entities from which our minds are built. They are the crucial components that determine who we are, the way we behave and the way in which other people view us.

1

A little miracle

When it came, the letter was brief. It was going to change my life, but I didn't know that at the time.

Would I please arrange to see this young man called David? He needed a neurological opinion. Unusually, he'd suffered a stroke in his thirties but thankfully had made an excellent physical recovery. In fact, within a few days he'd been walking and talking without difficulty, looking for all the world as if nothing had happened. However, it soon became apparent that all was not well.

David's behaviour had changed dramatically. Previously he'd been a very outgoing, highly motivated extrovert. But now he seemed entirely different. Although he'd returned to his job as a financial consultant quite quickly after his stroke, he was disinterested in the work that he'd only recently found so stimulating. He appeared bored. He lacked enthusiasm. Even when he was fired a few weeks later, he seemed quite indifferent. He couldn't be bothered to claim unemployment benefit and was now living with some close friends.

Although they had initially been happy to help out, they were fast becoming irritated with him. David had apparently been a highly gregarious person who always took the initiative at the centre of his social circle, but now he was quite 'hard work': disengaged, distant and aloof. Exasperatingly, he made very little contribution to the upkeep of the house. In fact, he did nothing all day, simply waiting for his friends to return in the evening to cook for and entertain him. His very inactivity made him annoying. He'd become a stranger to his closest acquaintances.

In his referral letter, his general practitioner went on to say that he'd started David on an antidepressant, thinking his behaviour was due to low mood, but the medication had no impact. The doctor was puzzled. Could we sort out why David had 'become so inactive, so dull'?

I winced. What did this doctor think neurologists did: turn boring people into interesting ones? Our specialty is usually the last point of referral for many unexplained symptoms, but this seemed absurd. Still, there were some features in this story that were distinctly unusual. First, it's relatively rare for a young person to suffer a stroke. Then there was the change in behaviour: this man was apparently so indifferent that he couldn't get around to claiming social security benefits. That was highly peculiar too. But if David wasn't depressed, what was wrong with him?

When he had suffered his stroke, David had been admitted to one of our sister hospitals, so I knew it should be possible for me to track down his investigations from my computer. A few clicks later and I was gazing at his MRI brain scans. At first, they looked pretty normal, but as I scrutinised them more closely I saw that he'd suffered not one stroke, but two. There they were: two very small holes, one on each side of the brain, almost symmetrically located, in the deep nuclei known as the basal ganglia.

This was definitely odd. Strokes don't normally occur as a brace like this. Usually, one side of the brain is affected because a stroke involves either the blockage of a blood vessel in the brain (referred to as an ischaemic stroke) or the bursting of a blood vessel (haemorrhage). To affect both sides of the brain means that the mechanism might involve blood clots that have originated from a distant location, e.g. the heart, and lodged in two different blood vessels, one on either side of the brain.

I worked through the list of blood and heart tests that David had undergone, but all had failed to reveal a cause. It wasn't surprising that he'd recovered physically very rapidly because these really were very small strokes. But why had he become so apathetic afterwards?

The story was curious. There were far too many unanswered questions. I decided we ought to see David soon.

~ o ~

I closed my eyes. It had been a very long clinic and the heat didn't help. From the open window, there was the throb of an air conditioning unit, which was no doubt keeping some other part of the hospital cool. We'd seen a lot of people that afternoon and I was feeling drained. To my disappointment, my odd patient hadn't turned up.

This was the outpatients clinic at the National Hospital for Neurology & Neurosurgery situated in London's Queen Square. Known affectionately for generations as 'the Square', this hospital had been home to some of the pioneers of neurology.[1] Matched perhaps only by the Salpêtrière, its rival establishment in Paris, during the nineteenth and twentieth centuries the Square had made enormous contributions to the understanding of brain disorders. And, by studying patients with neurological illnesses, the clinicians who worked there came to provide highly influential views on how human brains function.

That kind of historical tradition can, occasionally, weigh heavily on modern doctors. From time to time, we're reminded of just how impressive those pioneers were. They had diseases named after them and their systematic way of examining patients has been passed down, from generation to generation since. We still use the methods they developed to help make a diagnosis. However, in the modern health service we don't often have time to think about what we owe our illustrious colleagues from the past. We're simply too busy getting on with seeing the next case.

This afternoon was different. My potentially interesting patient hadn't turned up. It was a shame, but it wasn't my fault and maybe it was fortunate. I now had a bit more time to finish some letters on the patients I had seen.

Nonetheless, something niggled. Why hadn't he shown up?

I stepped out of the consulting room. The television in the waiting area was playing a news report of an anti-government protest in Syria as I walked past. I didn't think much of it. The Arab Spring had produced many surprises and this presumably was going to be yet another one. Nobody then could have foreseen how far-reaching the impact of these distant events was going to be, with Europe first welcoming Syrian refugees with open arms and then almost immediately shutting its doors to them as outsiders. But my mind that afternoon was focused on something far more parochial.

'This patient didn't turn up. Did he cancel?' I asked the receptionist.

'No! Do you want me to discharge him?' she replied eagerly. The National Health Service waiting lists are long and 'no show' patients like this just add to the time other people have to wait to be seen. I could understand why she was so keen to take him off our list.

'Oh no, do you have his phone number, please?'

I don't know what made me ask, but I was now far too interested in his story. Somehow, I had been sucked into the tale of this man who had suddenly become 'boring'!

To my surprise, the telephone rang only twice before it was picked up.

'Is that David?' I asked

'Yeah,' came the reply. It was a matter of fact, prolonged drawl, a disinterested voice, one that seemed accustomed to saying very little, somehow minimalist without even trying.

'David, I'm calling from the hospital. I'm the neurologist you were supposed to see. You were due to be at my clinic this afternoon. Any reason why you couldn't make it?'

There was a long pause.

'Ahhh . . . Sorry about that . . . It just seemed like a long way to go.'

'I see. Look, David, I think it's important we see you. Are you free tomorrow morning?'

Another long silence.

'OK. I can do that.'

'Are you sure you'll come along?' I asked. 'I'll be waiting to see just you.'

'Okay, where and when?' came the response after a few seconds.

It was grudging. I could almost see the shrug of the shoulders, as if I was somehow infringing on his precious time. He didn't seem at all excited. I found that mildly irritating. Even though we try not to show it, doctors suffer the vanity of thinking patients might need us – might even want to see us. Not this one, it seemed.

~ o ~

The next morning, from my office window I looked down into the large gardens which occupy the centre of Queen Square. Over the canopy, I could just glimpse the slate-grey statue of Queen Charlotte. There she stood, slightly awkward, her right hand outstretched but empty, clutching at the air. It had once held a mace but revellers had extricated it a long time ago. In

Victorian times, because there was no original inscription, the statue had been considered to be a likeness of Queen Anne and, as a result, the square had been named in her honour. Much later, closer scrutiny suggested that Anne was very unlikely to be the monarch who had inspired this figure. Instead, the statue was of Charlotte of Mecklenburg-Strelitz, who, at the age of seventeen, had travelled from a small northern German duchy to marry the young King George III.

When she arrived in London, Charlotte apparently spoke no English. She was considered ugly by the court, deemed to have very little going for her. Indeed, some claimed that her facial features were far too foreign, reflecting a Moorish line of ancestry that can be traced back to a branch of the Portuguese royal family. Whatever the truth, Charlotte found it extremely difficult to connect with the English, but she seems to have been determined to try. She learned her new language rapidly and excelled at it, but she never lost her German accent. Her voice proved to be her ultimate weakness and, apparently, she never did really feel like she belonged.

I could relate to that. When I'd arrived in the United Kingdom in 1968 as a child, I recall feeling extremely proud that I spoke English well, having been taught it from a very early age in what was then East Pakistan. Much later, at the age of ten, after I had excitedly recorded my own voice on my father's new Philips reel-to-reel tape recorder, I listened with shock to speech that I didn't recognise. It was delivered with a cadence that seemed almost alien to me. I had, I surmised with some astonishment, a foreign accent. No matter how distinguished I considered my English to be, I was clearly deluded. This accent, together with the small matter of my brown skin, was what marked me out as being different. I resolved that I had to change, otherwise I would keep sounding like a foreigner. For the next three years, I listened diligently to BBC radio, practising my pronunciation and religiously recording my own voice to monitor if my accent was improving. Like Charlotte, I wanted to fit in.

Years after her arrival, when King George suffered his first bout of madness, attributed by some to the metabolic illness known as porphyria,[2] by others to arsenic or bipolar disease,[3] he was tended to by Francis Willis, a physician whose house was apparently in Queen Square. Willis's techniques were part bullying and intimidation, part ingratiation. He had the King restrained, sometimes in a straitjacket, sometimes gagged.[4] The Queen, distraught but keen to tend to her husband, supposedly found another house in the Square where she kept provisions for him. Nowadays, this is the location of a pub, appropriately called the Queen's Larder and frequented by many neurologists.

Eventually, George III's condition improved, probably due to natural recovery, but Willis took the credit and was lauded for his achievements. Although the King was to suffer further episodes, the link between brain disorder and Queen Square had been initiated. One certainly cannot escape history there. It pervades the place, cloaking its modern neurological occupants with their own peculiar culture of belonging – in this case to a long and rich tradition, to a legacy that is hard to ignore.

David did eventually turn up that morning. It was a surprise to me. I was giving up hope when he arrived forty minutes late. 'Trouble with the buses' apparently. I didn't quite believe him, but I put that aside. He wasn't what I had imagined.

In front of me sat a short, wiry man with a narrow face, framed by thick, black-rimmed spectacles through which he peered with deep disinterest. Dark brown, greasy hair slid over a sweaty forehead, and a ragged moustache, flecked with crumbs of cereal (perhaps from breakfast that morning), sat uncomfortably over his thin lips. A thicket of hair, surprisingly curly and rusty in colour, covered his cheeks and chin, more coarse shrubbery than a beard. He sported a crumpled grey shirt that looked like it had clung to him for several days, and possibly nights, carrying food stains of a multitude of vintages. David looked so much older than his years. It seemed that he wasn't looking after himself.

'Thanks for coming along,' I said, wide eyed and probably giving away too much of my concern about the state he was in.

'Sure, yeah.'

'I understand you were in hospital earlier this year. What are your biggest problems now?'

'Problems? I don't have any problems.' He shrugged. That movement of the shoulders made me smile. I had imagined exactly such a response when I'd spoken to him over the phone. David's face, by contrast, was far less expressive, but he looked as if he was unimpressed with my opening gambit. In fact, he seemed extremely unimpressed with me generally.

'Well, your doctor wrote to ask me to see you.'

'Did he?' He pushed his heavy spectacles up the bridge of his nose.

'How are you after the stroke?'

'Fine. I got better pretty quickly.'

'So I understand. You don't have any weakness of your arms or legs?'

'No.'

'You don't find anything more difficult?'

'No'

'What about your job?'

'What about it?'

I was clearly not getting very far; each of my questions was resolutely being batted back with a straight face. David was going to be hard work. 'How did you find it when you got back to your job?' I asked, trying to give him room to expand.

'Fine,' he said, not giving me an inch with his riposte.

28

'Did your employers think the same?'

This was now very direct from me, but I got the distinct feeling that I was going to have to be direct with David. He wasn't going to tell me anything useful of his own accord.

'What do you mean?' he asked coyly, his eyes narrowing in apparent bemusement.

'Are you still working?' I asked very slowly and deliberately. Now the lips tightened, but only for a moment, before we were back to his expressionless face.

'No, I'm not.'

I gave him time to say more, but there was no more. It was like talking to a brusque teenager. This was fast becoming a very staccato conversation.

'Why's that?'

'Things weren't working out,' he responded vaguely.

'Not working out?'

'Yeah, not working out.'

To extract information from a complete stranger is an art that doctors have to learn. Neurologists, in particular, pride themselves on their ability to draw out the details so that a diagnosis becomes clear even before they've examined a patient. In medical school, students are taught to use open-ended questions, but most experienced doctors work out that they have to direct the questions, depending on the quality of the responses from the patient. David, however, was exposing just how difficult the process can sometimes be and I could feel myself getting frustrated by his diffidence. But I wondered if his brevity in response might in fact be part of his condition, so I tried a softer, less direct line.

'Look, I'm only trying to find out what's happened to you.'

I said no more and allowed a sepulchral silence to fill the room. Outside, a truck rumbled by; the noise amplified by the lack of words between us.

'They fired me,' he said eventually. His unkempt moustache quivered a little.

Again, I decided to hold back to see if he would say more, but he didn't. It was almost as if he needed to be prompted to give the next bit of information. 'What did you do?' I asked very deliberately.

'Nothing. That's the point. I couldn't be bothered to do anything, so they fired me. They sacked me just like that.'

'That sounds terrible,' I responded, trying to be sympathetic. 'You said that you couldn't be bothered. You mean when you went back to work after the stroke?'

'That's right. I didn't care. I guess it showed.'

David was quite clear that he hadn't been feeling low or depressed. He just couldn't be bothered. No, he wasn't worried about the future and he definitely hadn't thought about getting a new job. He shrugged his shoulders. We were going down the wrong channels, again. I wondered whether it might be me. Was my irritation showing? Or perhaps this really was part of David's condition, the reason why his friends were becoming frustrated with him too.

'How is it, living with your friends?' I asked, trying to change the subject once more.

Apparently, it was fantastic. They'd been really great to him and were real mates. He could count on them but surprisingly he didn't seem either genuinely enthusiastic or grateful for what they'd done. He'd been with them for four months and he was quite honest in saying that he had moved in with his friends because he could no longer afford to pay the rent on his own flat. But when I asked him whether unemployment benefit would have covered this, he responded eventually by saying he couldn't be bothered with the hassle of filling in all those forms. No, he hadn't gone to an employment centre either because, as he put it, he 'couldn't be arsed'.

'What's going to happen, then?'

'How do you mean?' he asked indifferently.

'Well, your friends seem really kind, but how long can you stay with them for?'

'Oh, for as long as I like – I think. They haven't said anything about how long I can stay. They're alright with having me about,' he added with a grin.

I decided to move on and asked him what he was doing during the day. Apparently, he spent much of the time sitting on his own, doing nothing. Previously, he would have listened to music or read novels he explained.

'Any reason why you're not doing any of those things now?' I asked.

'Well, I'd like to listen to music, but I'd have to put my music system together and I just haven't taken the trouble to do that.'

'Oh, how long would it take to put your hi-fi system together then?'

Another pause, but I allowed it to come between us. David looked bemused again.

'Five minutes. I'd have to unpack it, stick the leads into the speakers, plug it in, you know. But I can't be bothered.'

This was revelatory. David had lost his job but wasn't apparently unhappy. He didn't seem to think his situation was hopeless despite the fact that he wasn't making any attempt to find a new job. He was living with friends – actually, living off them – and, as I found out later when I spoke to them by telephone, in his own way he'd become a bit of a stranger to them: dishevelled, no longer caring about his appearance, not even bothering to have a shower unless they prompted him. The man who had been at the centre of his social circle – the insider – had been transformed into someone who was quite alien to them. Yet, despite all this, he was quite remarkably indifferent to his new circumstances. He had developed pathological apathy, all of a sudden.

As David left the consulting room, I wondered if I would ever see him again. My mantra to medical students is that, whatever we do as doctors, we should always try to make a patient feel better leaving the clinic than when they came to see us. This means connecting with the person as an individual, but in this case I didn't feel like that had happened. At best, David had been indifferent about the consultation. At worst, from his point of view, he hadn't really got anything out of it. But had I?

~ o ~

Why couldn't David be motivated to do anything? Why wasn't he bothered? In nineteenth-century Russian literature, Ivan Goncharov in his novel *Oblomov* created a seemingly similar character. Oblomov is a young landowner who inherits an estate but is completely incapable of taking any action or decision. For long, languorous stretches of the novel he lies in his bed, moving only occasionally to sit on his chair. He is the paradigm of inertia. David was, in many ways, a mirror image of Oblomov. He was apathetic, indifferent and seemingly not motivated to do anything. But there was one important difference. David had changed character suddenly, after his unusual strokes, whereas apathy was a lifelong trait in Oblomov.

So why didn't David do anything? He certainly wasn't depressed, so that wasn't the reason. The more I thought about it, the more shocking his inactive predicament seemed to me. Could it be that the location of his brain damage was telling me something important about normal motivation in all of us? Had his strokes extinguished his motivation because they had effectively taken out some of the circuitry that normally makes healthy people motivated? Had David become apathetic because these regions weren't working properly? I needed to find out more.

David's strokes had affected the basal ganglia. These nuclei lie deep in the brain (**Fig. 1**). They are ancient structures, present

32

throughout the vertebrates (backboned animals) all the way back to the humble lamprey, a species of jawless fish which emerged 360 million years ago and still survives to this day. But, although the basal ganglia have a long ancestry in evolution, as Kinnier Wilson, one of Queen Square's famous neurologists put it in his lectures to the Royal College of Physicians in 1925, they are like the dark basement of the brain.[5] At the time of Wilson's lectures, people certainly understood very little about the functions of the basal ganglia. We now know far more, but still have a long way to go to figure out all that goes on, down in the basement.

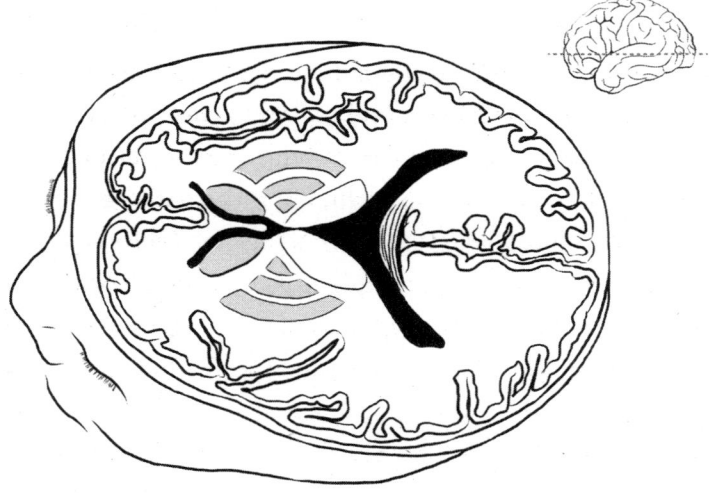

Fig. 1 | The basal ganglia. These structures (shaded grey) are located deep in the brain, as shown in this section through the brain. In David, small parts of the basal ganglia had been affected by his strokes on both sides of the brain. The image of the brain on the right shows the level at which this section is taken.

Although they lie deep from the surface of the brain, the basal ganglia are highly connected to the overlying cerebral cortex. Different areas of the cortex send connections to different parts of the basal ganglia, which in turn send projections back to the cortical regions, including areas in the frontal lobe involved in the control of movement.[6–8] These massive loops of connections between the basal ganglia and the frontal cortex have been found to be critical for many functions, including the initiation and invigoration of movement.[9–11] Indeed, for many decades the basal ganglia have been associated with the control of actions, largely because of clinical observations of abnormal movements resulting from their damage: such as those that occur in Parkinson's disease, Huntington's disease or even the rare illness named after Kinnier Wilson himself – Wilson's disease.

In all these conditions, there are profound changes in movement. In Parkinson's there might be a slowing of action, or a tremor or rigidity; in Huntington's there can be unwanted, sudden thrusts of the head or flinging of limbs; and in Wilson's there might be strange, awkward posturing of the limbs termed dystonia. But although early work on the basal ganglia focused on their role in the control of movement – and how this goes awry in brain diseases – it has since become clear that they do far more.

Most of our modern insights into the functions of the basal ganglia have come not from studies of human diseases, but rather from fundamental neuroscience. Particularly interesting is the work of Canadian physiologist Gordon Mogenson. Neither animals nor humans make movements without good reason he argued.[12] There needs to be a 'motivation for action'. Mogenson proposed that animals move to satisfy 'internal' primary motives: the need for food, water, to get warmer or cooler, and for sex. This is necessary to survive and to secure the species.

For Mogenson, the basal ganglia are a crucial bridge between motivation and the initiation of action. Think for a moment about the sensation of hunger. We experience it several times a day, but what is it? It's an urge to find food in order to quell the sensation we call hunger. But why does that sensation arise just a few hours after our last meal? Most of us are not going to starve if we don't have food every few hours. Yet the drive to seek calories arises irresistibly in our minds, producing a compelling urge to leave what we're doing and focus instead on finding food. Hunger is an irresistible motivator that we all succumb to.

Mogenson considered two parts of the basal ganglia to play a key role in effectively *converting motivation to action*. One of these, known as the nucleus accumbens, was intact in David, but the other critical area known as the ventral pallidum had been damaged by his strokes. A large number of investigations in animals, mostly rodents, has now revealed that both these regions are necessary for motivated behaviour.[13,14] When the nucleus accumbens or the ventral pallidum become dysfunctional, rats become less motivated to make an effort. Whereas normally they might be prepared to work hard to obtain food that they really like, when these areas of the basal ganglia are disrupted, the animals appear quite indifferent. They become disinterested in putting in effort to acquire rewarding outcomes such as food that they previously prized highly.

The more I dived into these studies, the more I began to see similarities to what had happened to David. He really couldn't be bothered to put in effort to receive rewarding outcomes – like getting paid for doing his job or putting his music system together to listen to his favourite tracks. How could we find out whether he was sensitive to rewards? I had an idea.

~ o ~

We had finally got David into the research centre at the Institute of Cognitive Neuroscience located opposite the hospital in Queen Square. He'd missed two of his previous appointments. To ensure he made it, we had to resort to booking a taxi to pick him up. He arrived even more dishevelled than before. His clothes looked like they hadn't been washed for weeks, and the same might also have been true for David himself. The moustache and beard were now overgrown, a thicket of ragged curls and twisting russet tendrils. His fingernails were enormously long, some splintered, all discoloured by streaks of dirt underneath them. Things were obviously getting worse.

'I'm really sorry for not getting here earlier. I just found it difficult to get my act together,' he said, wiping his spectacles with a dirty handkerchief.

'Don't worry,' I responded. 'Thanks for agreeing to take part in this research. How have you been?'

David looked rather sheepishly to the floor, his thick-rimmed glasses slipping sightly down his nose. 'Oh fine, thanks,' he responded.

'Are you sure?' I asked gently.

David hesitated, his face still not giving away his emotions.

Then he said, 'I've actually been having a bit of trouble with my mates, if you must know.' There was another long pause before he explained that his friends had been getting annoyed with him. He cleaned or did the shopping, but only when they asked him. Nothing happened without them prompting him. Now they were giving up even asking him to do things. He realised his shortcomings, but he just couldn't be bothered if they didn't remind him. He never really thought about doing anything, he explained, let alone helping with the domestic chores.

David was telling us something interesting: he needed to be nudged to achieve everyday tasks. He lacked the ability to motivate himself to do these things. He didn't have *motivation to act*, as Mogenson might say.

'Is it alright between you and your friends otherwise? Are you getting on with them?' I asked.

'Not really. They don't talk to me much. I get the feeling they avoid me. They don't eat with me anymore. Sometimes I don't feel like I really belong there . . . but I'm sure it'll be alright if

I do a bit more around the house,' he said, convincing neither of us.

David was becoming alienated from his friends who, not so long ago, had considered him as one of their own. His reluctance to do anything without prompting had frustrated the very people who really liked him. Within a few months, it had led to them not engaging with him, just as he was not connecting with them. It was that easy to fall from the comfort of belonging to the 'in-group', as social psychologists would put it, to losing a sense of belonging and becoming part of the 'out-group'.

Motivation to behave in the way that is expected of a group member is a key part of belonging. We have expectations about how 'engaged' our friends, family members and colleagues should be with us. And they also have presumptions about how motivated we should be towards keeping connections with them. They have expectations about the contributions we make to the group.

The level of commitment required to keep our membership differs across different types of social networks. It depends partly on the elasticity of tolerance other members are willing to stretch to, as well as the longevity of the group. Some communities can have remarkable patience with the unsociable behaviour of their prodigal sons and daughters, who might be fortunate enough to enjoy the benefits of a strong relationship built over many years. The same degree of tolerance might not be afforded by people we've only just got to know. As a result, group memberships can lapse, our social networks can collapse, if we don't put in the effort to maintain relationships.

Part of the quality of that endeavour is often judged by the yardstick of communal responsibility. Within a household this might be measured by the willingness to keep the kitchen and bathroom clean, restock the fridge, cook for others in the house, make an effort in conversation, or even suggest social activities.

David did none of these things. Then there's the issue of hygiene, the cleanliness of the clothes one wears, and the attention one pays to grooming. Demeanour matters. Again, David's evident self-neglect wasn't helping his cause.

All these factors, often tallied up subconsciously but sometimes scored explicitly, provide a metric of motivation by which we are judged. Lack of commitment to contributing to the overall good of the group, in any way, is soon noticed and, as some of us have undoubtedly experienced, there are consequences. Robin Dunbar, an Oxford anthropologist and evolutionary psychologist, and his colleagues observe that free-riders (those who take the benefits from being in a community but do not contribute) challenge the mutual trust within a group.[15] According to them, social groups originally evolved because of the many benefits obtained from being together, including mutual protection, hunting, and the ability to share food and knowledge. But belonging entails 'signing up' to unwritten social contracts, which involve trusting other members of the group and honouring social obligations that are embedded in a community. A key aspect of this is a willingness to make an effort or contribute without necessarily expecting an immediate return on that invest-ment. David wasn't doing any of that of his own volition.

Failure to meet the social expectations of motivated behaviour tests the forbearance of people within a social group. Nevertheless, it is still possible to be allowed to remain within it if our behav-iour reverts to what is expected – the norms for that group. But for many individuals with a brain disorder that is often not possible. Under such circumstances, a group member can effec-tively be ostracised. There is a cost attached to group affiliation, as David was finding out.

I took him to the testing room. First, we performed some screening cognitive tests, assessing his perception, attention, memory, language, visuospatial skills and aspects of frontal lobe function.

David passed all of these without any difficulty. Next, we moved on to the main reason I had brought him back to see us.

Around that time, in my research group, we had developed a new test to detect impulsive, risky decision-making for rewards. It had proved to be very useful in measuring how willing people were to take risks if the reward was sufficiently attractive for them. Although we hadn't thought of it before, I now wondered whether we might also use this new task to investigate the converse. Could it reveal when someone was not incentivised by the prospect of rewards?

I sat David in front of a computer screen.

'So, what do I do?'

'Do you see the traffic light on the left-hand side of the screen?' I asked, pointing to it.

'Sure,' he said.

'In this test, I want you to look at the traffic light. Now it's red but it will turn amber then green. When it goes green you have to move your eyes as quickly as possible to look at the white square on the other side of the screen.'

'OK, sounds simple.' David nodded.

'The quicker you move your eyes to the square, the bigger the reward you win. And we measure where you're looking using this fancy camera,' I said, pointing to a bit of expensive kit that allows us to measure where the eyes are looking.

'So, the faster I react the more I win? What exactly do I win?' he asked inquisitively.

At least I had piqued his interest I thought.

'Well, in this test it's actually money! You see, the faster you go the bigger the reward in terms of the money you make. The computer tells you how much you make on each trial.'

'Sounds easy!'

'Well, there are a few things you need to know about this game, David. If you make an eye movement before the traffic light turns green there's a penalty. You lose a small amount of money. On the other hand, if you go late after the amber turns to green, you make very little money. So, it's more challenging than you think!'

'Okay, I'll give it a go,' he said.

Most people find it fun to do this test. It's a game for them and they're keen to accumulate as much money as possible. But it's actually more challenging than they initially think. To win a big reward they soon find out that they have to anticipate when the amber light will turn green. However, to make life difficult, the exact duration of the amber light isn't constant, but varies from trial to trial. So, it is never possible to predict with absolute certainty when it will turn green. On the other hand, if people just wait for the green light, they end up winning very little. To do well on this test – to earn the most amount of money – you have to take some risk.

Ideally, your eyes should start moving exactly as the traffic light turns green. However, because it takes 200 milliseconds (a fifth of a second) from wishing to start the movement to the eyes actually moving, you have to make the decision to shift your eyes *while the traffic light is still amber.* And because you never know exactly how long the amber light lasts, you have to take a risk about whether to initiate the eye movement while it is still on. If you take the risk, of course, on some occasions, you'll go before amber turns green and incur the small flat, fixed penalty. However, on other occasions you'll hit it right: your eyes will start moving just as the traffic light goes green. Then, you can make a lot of money because your eyes can shift to the box on the other side of the screen before the reward level falls very much.

In short, if people are incentivised to obtain a high reward, they start to take risks and plan to make their eye movement during the amber light – not wait until the light turns green. Individuals vary in their willingness to take such risks but everybody does to some extent, particularly when the amber light duration is long.

David, however, was completely different in how he approached the task. He could understand it and perform it, but he would just wait until the traffic light turned green before he made his eye movement. He never anticipated. He never initiated movements before the green light came on.[16] As a result, compared to other people of his age, he made almost nothing in terms of the total reward earned. Unsurprisingly, he was indifferent.

'Yeah, that was interesting,' he said, clearly not at all curious or perturbed that he'd done so badly. He sat with his arms folded, waiting.

It had been very instructive for us to observe how he performed this task. It showed us that David was capable of taking actions *when prompted* by an external signal – by the traffic light turning green. But he would not initiate those actions himself in order to maximise his rewards. The results of several other assessments we made also led us to conclude that an important feature of David's apathy was that he was not incentivised, of his own volition, to obtain rewards. He had blunted – possessed very little – sensitivity to reward. He couldn't be bothered to initiate actions himself because the outcomes didn't matter to him. Instead, he just waited until prompted to make the action by the green light coming on.

This was analogous to what happened to David in real life. He'd do nothing, or very little, of his own volition, but he was quite happy to clean up, go shopping or even cook if he was asked to do so by his now increasingly frustrated housemates. Given that on the traffic light test he seemed to be insensitive to rewards that would normally motivate most of us, I began to wonder whether there might be a way in which we could attempt to restore reward sensitivity in David.

~ o ~

41

The brain chemical dopamine has had a long association with pleasure and reward. In fact, many research studies in the 1960s and 70s began to lead to the hypothesis that dopamine is the brain's 'pleasure chemical'. Our views about what dopamine does have changed quite a lot since then. Most researchers now believe that dopamine acts as a key chemical *to motivate us to seek rewards*.[17] Rather than giving rise to pleasure directly, it promotes us to be motivated to search for pleasure or reward. Perhaps most relevant for David is the fact that dopamine is also a key brain chemical for the basal ganglia. When parts of the basal ganglia are depleted of dopamine, people develop the motor features of Parkinson's disease, a condition that was untreatable until researchers found a way to boost the level of dopamine in the brain. Some of those early studies are beautifully recorded by Oliver Sacks in his book *Awakenings*.[18]

Neurologists such as Sacks used the precursor of dopamine, laevodopa, combined with a drug that would prevent its breakdown in the blood if it was taken by mouth. This combination can have astonishingly positive effects on improving movement in people with features of Parkinson's disease – seemingly 'awakening' them by increasing the amount of dopamine produced in the brain. But could such a drug be of any use for someone like David? He didn't have a problem with moving. His difficulty was in *motivating himself to move*, to act, to do anything. He didn't seem to initiate any actions unless he was prompted to do so.

Work in animals suggested that it might be worth trying laevodopa. Researchers have found that when rats have low levels of dopamine, even if they are able to make movements, they seem less motivated to invest physical effort for rewards.[13] I discussed this with David. He was, as ever, indifferent. Initially, there was the inevitable shrug of the shoulders, but eventually he decided to try it.

'Alright, I'll give it a go if you think it might help,' he said, showing not a flicker of excitement at the prospect that there might be some way to improve his current state.

'I can't guarantee it will, David, but I think it's worth seeing how you do on this medicine.'

As he walked off with his box of tablets in his hand, I couldn't help but wonder whether he would even be bothered to take the drugs.

Slowly, over several weeks, we increased the dose of laevodopa that David took, making sure, as best we could through phone call prompts, that he was actually taking the medication. Three months later, although he said he felt better, he seemed unchanged. Furthermore, when we asked him to perform the traffic light test again, there was barely any difference. He remained just as insensitive to potential rewards as he had been before, simply waiting until the traffic light turned green before he moved his eyes.

'Yeah, that was good,' he said when he completed the test.

But, clearly, laevodopa hadn't had any noticeable impact on his apathy or his sensitivity to rewards. This was very disappointing, so we decided to take him off the medication slowly. Before we gave up on this approach completely, however, one other thought occurred to me. Many Parkinson's disease patients are now treated not only with laevodopa but also with drugs that mimic dopamine by locking on to specific receptors on the nerve cells that bind dopamine. These dopamine receptor agonists – often simply referred to as dopamine agonists – can be highly effective in treating the movement impairments of Parkinson's disease.

Over the last few years, however, we have begun to realise that sometimes, at higher doses, dopamine agonists can tip some patients into making impulsive, rash decisions such as gambling too much money, going on shopping sprees for things they don't need or becoming hyper-sexual. In short, on these drugs, some people can become too highly incentivised to seek rewarding outcomes. One of our patients, for example, managed to spend over £10,000 on online betting in a day. When the dose of dopamine agonist is reduced, these behaviours can resolve, so they appear to be directly induced by activation of dopamine receptors within the brain.

Similar impulsivity occurs far less often with laevodopa, perhaps because laevodopa increases dopamine produced by nerve cells in the brain in a general manner, whereas dopamine agonists directly stimulate specific dopamine receptors on neurons. Some of these receptors might be particularly important for initiating actions. If dopamine agonists can also turn Parkinson's patients towards becoming impulsive and taking risks, might they be effective in treating David's apathy by making him more sensitive to rewards and thereby energise him to initiate actions? I put this idea to him and again, although nonplussed, he decided that he'd like to give it a try.

'Sure, if you think this'll be any better than the last lot of pills,' he said.

Cautiously, we started him on a dopamine agonist called ropinirole. Neurologists have extensive experience of using this drug in patients with Parkinson's disease, but we had no idea what it would do for David and his lack of motivation. Very slowly, we increased the dose of ropinirole, again calling him regularly to check that he was taking the medication, and that he wasn't experiencing any side effects such as the impulsive behaviour that can occur in some Parkinson's patients on dopamine agonists. Three months after starting the treatment we managed to persuade him to see us again in person to check on how he was doing.

~ o ~

The receptionist at our research centre called me to say that David had arrived. When I went down to meet him and looked around the waiting room I couldn't see him. There was only some guy in a smart suit and a young woman there.

'I thought you said David was here?' I asked the receptionist.

'He is. That's him over there,' she replied, nodding her head in the direction of the man in the corner.

I turned my gaze towards him. As I took a closer look, it was clear that the receptionist was right. This was in fact David, but barely recognisable. The greasy hair which had previously covered much of his forehead had been elaborately cut and combed back to a quiff. The moustache, normally lined with crumbs, and the untended hedgerow of a beard had been dispatched. In its stead, a clean-shaven visage proudly basked in the daylight. I peered at David in disbelief. He looked so young. His outfit was shocking to me, too. A crisply ironed white shirt, adorned with an under-stated lavender tie, lay under an immaculate navy blue suit. Sporting a pair of polished brogues, David sat one leg crossed over the other, a leather briefcase by his side. As he rose to greet me, a flash of silver cuff links capped off his startling new look.

'Hello!' he exclaimed.

It was my turn to be slow in responding.

'You well, doctor? You look a bit pale if you don't mind me saying so,' he said cheekily, smiling.

'I'm fine, quite fine,' I said. 'Tell me . . . tell me what's been happening to you since we last met.'

'I'm different!' he explained, spreading his arms wide open.

'You certainly are,' I observed after another pause. 'Very different.'

I looked at him closely again, taking in more details of the new David. His fingernails had been neatly trimmed. No dirt under those, I thought, and no food stains anywhere either.

'Sure you're OK?' he asked again, smiling broadly.

'Yes, I'm quite sure, thanks.'

'Oh good. You'll be pleased to know I got myself a job, doctor.'

45

'A job? How did you do that?' I blurted out.

'Well, I answered an advertisement in the papers. Simple as that. Of course, I had to get my CV together, spruce up a bit, look the part.' He laughed, pointing at his suit.

'What kind of job is it?'

'It's in a financial services firm, like the one I was in before, and it's working out really well. The pay's good, so I left my friends' place and rented a flat. The last few weeks have been fantastic.'

'Really!' I said slowly.

'Yeah, and guess what, doctor? I've even met someone. She's brilliant! We've been going out for the last month. I'm really happy.' David was beaming.

I had not seen so much expression in his face in any of our previous discussions. This also felt like the first natural conversation between us. He was even volunteering information spontaneously, without me having to drag it out of him. I considered what he had said about his new job and flat. But the news that David had a girlfriend, when he'd previously barely gone out and had been so dishevelled was, to say the least, perhaps the most shocking of his revelations that afternoon. His life seemed to have been transformed. Of course, it could have been pure chance, but that hardly seemed likely given how long he'd been apathetic.

We got him to do the traffic light test again and this time he behaved differently.[16] Like normal people, he started to take risks. He started to anticipate and initiate eye movements before the traffic lights turned green. The amount he won increased substantially compared to his previous attempts on this test. So, it appeared that ropinirole, the drug that clung to and directly activated dopamine receptors in the brain, had made a huge difference. It had incentivised him to seek rewards on our test and, far more importantly, in his life too. David was now a seemingly highly motivated individual.

'This is amazing! It's great to see you doing so . . . well,' I said, still not quite believing the change in him.

'Thanks, doc. It's a little miracle!'

'You might well be right, David.' I nodded.

'Can't hang around, though, as I'll be late for drinks with my old housemates. Things are a lot better on that front too by the way. Before I moved to my new flat, they were really impressed that I was doing more of the cleaning, preparing a meal before they got back home, probably also that I was looking a lot tidier myself!'

'Really?'

'Yeah, really; those pills might have done the trick. But I've got to go! See you soon and thanks for everything.'

And off he went. He never would have worried or bothered about being late, let alone turning up to meetings with people before. Now he was in a rush. I was quietly elated. This was a day to celebrate a very unusual achievement. Had we really found a cure for apathy?

~ o ~

David proved to be a turning point in my life and research. He was living proof that it was possible to turn someone who had suddenly developed pathological inertia back into being a highly motivated individual, capable of leading an independent and rewarding life. Motivation is a key attribute of human brains and of being human. It's a brain 'function' we usually take for granted, but it shapes our identities. We have all come across people who seem highly motivated – at school, in a sports team, at college, in the workplace – who often also seem to be highly successful. Meeting David made me aware of how easy it is for someone who is at the centre of his social group suddenly to become an outsider and to lose their sense of belonging, even to close friends, simply because of a change in their level of motivation.

David was, of course, an extremely rare person. Strokes like his, which selectively disrupt the ventral pallidum of the basal ganglia, are extremely rare. But it turns out that apathy is an extremely common syndrome that occurs in many other brain disorders such as Alzheimer's disease, vascular dementia and Parkinson's disease (which we now know can have manifestations that are not just motor).[19] Roughly a third of patients with these diseases develop pathological apathy. These conditions are all neurodegenerative disorders where parts of the brain slowly begin to shrink, losing nerve cells over time. Until recently we didn't really have a good idea about why such diverse diseases might lead to the same syndrome of apathy. The insights we gained from seeing and testing David in our research provided an important answer.

Although his apathy was due to two tiny strokes, these happened to be very strategically located. Despite being extremely small, they had rendered dysfunctional a key brain circuit that connects the basal ganglia to frontal regions of the brain. And in so doing they had disrupted what Gordon Mogenson had termed 'motivation to action': how goals that are highly rewarding trigger actions which allow us to attain them.[12] It turns out that the same circuitry is also disrupted in neurodegenerative conditions that can be associated with apathy. Using brain scanning we were able to reveal that the circuit that was dysfunctional in David is also disrupted in apathetic people who have vascular dementia.[20]

Whether this same circuit has implications for understanding the milder forms of apathy that occur in healthy individuals is an intriguing question. Until I met David, I had thought that there are some people who are highly motivated and others who are simply lazy. David's strokes revealed that there might be a biological mechanism that could be crucial to how motivated a person might be. When part of this circuit is taken out, motivation can be extinguished.

David made me wonder whether we might be able to detect differences in brain activity between highly motivated people and apathetic people who are otherwise healthy. Remarkably, we found that there are indeed such variations – located in the basal ganglia and the frontal cortex – when healthy volunteers were asked to decide whether making an action is worth the effort to obtain a reward.[21]

Of course, there might be many reasons why some people lack motivation – not simply because they're not incentivised by rewarding outcomes, but perhaps because they have become depressed. But our brain scan findings in healthy individuals suggest that some of the differences in motivation we observe in people might be explained specifically by alterations in the brain circuit related to motivation, not mood. There might, therefore, be a biological mechanism that accounts for some of the variability in levels of motivation we encounter across our friends, family members and colleagues.

In David's case, motivation was extinguished suddenly because strokes occur abruptly. However, in neurodegenerative disorders like Alzheimer's or vascular dementia, it can happen very slowly, creeping up on both patients and their families over many years. In my clinic, I see many wives or husbands of patients with such neurodegenerative diseases who say that they understand why their partner can't remember information. But what really irritates them is that their spouse will no longer help with the washing-up or take the bins out, just as David's friends complained about his everyday apathy.

I explain that this lack of motivation is part of the disease, it's not the patient's fault and, most importantly, they are not wilfully being inactive. It's just that their brains no longer have the ability to generate motivation, and they might have very little insight about the fact that they're so inert. This can seem perplexing at first. But once people appreciate that the apathy is caused by

49

the disease affecting parts of the brain that normally motivate us to action, they are more willing to be tolerant.

Of course, what they still yearn – long – for is their partner to return to their previous selves. We found that David's pathological apathy was due to the fact that he had lost sensitivity to rewards. He didn't do anything because he didn't find any particular behaviour to be very rewarding, so there was no motivation to act. Dopamine turned out to be a key chemical that ramped up his sensitivity to rewards and restored his motivation to take action. Exciting new work in neurodegenerative conditions such as Parkinson's disease and Alzheimer's disease has begun to show that drugs that increase dopamine levels can help treat apathy in people with these conditions too.[22] However, dopamine is unlikely to be the only chemical in the brain that affects motivation.

We're just at the beginning of understanding apathy in neurological diseases, with several research teams around the world making rapid progress in this field. But my personal foray into apathy all started with meeting David. What people like him have shown us is how essential motivation is in constructing our personal and social identities, who we are and how others perceive – and evaluate – us.

2

The man who ran out of words

'This has been cool, not like any other medical clinic I've been to,' she said in her strong south London accent.

'I'm glad you've found it interesting.'

'But how come there are all these conditions I haven't even read about in our textbooks?' she asked curiously.

'Well, I guess it depends upon the textbook you read. Anyway, I thought you all get your information from the internet nowadays. Do medical students still read books?' I asked, smiling at Amina.

I had met her once before on one of our teaching rounds, and she had struck me then as someone who wasn't shy of speaking her mind. In fact, quite the opposite. I believe the modern term for someone like her is an 'over-sharer'. It was Amina's turn to sit in on my clinic as she went through her four-week Neurology rotation.

'Some of us do still read books,' she said with a smirk, raising her eyebrows. 'But only if we get advice on which books are worth reading. So, what should I read?' she asked, spreading her delicate, dark hands wide apart for emphasis.

I nodded, impressed by her enthusiasm. It had been a long

afternoon. Now it was twenty to six and my last patient still had not arrived. Two extras had been overbooked earlier into the clinic because they needed to be seen relatively urgently. We had been able to deal with them quite quickly, but I had also been teaching Amina which, although fun, had also proved to be rather discursive.

Without prompting, Amina had explained that her family had escaped from Somalia when she was a young child. They had been refugees from the civil war and had survived a very difficult journey to the UK. She didn't really remember anything about Somalia. She had never been back. All she really knew was London, but it had been a real struggle growing up here when she was young. I probably didn't have any real idea of what it was like, she assumed (and I didn't correct her), but it had been hard to fit in, very hard. But she had been so pleased to get a place at the medical school a few years ago. She was loving it. It had been a liberating experience, one that freed her from the cultural expectations of women in her family while exposing her to all this wonderful learning.

'Well, if there's one book you might look at, it's this one,' I said as I scribbled down the details on a Post-it note. 'You'll find it in the library . . .' I was finishing my sentence when she interrupted me.

'And before you say it. Yes, I do know where the library is,' she responded, in a half-mocking tone, smiling broadly.

'I would never have thought otherwise, Amina. Our last patient hasn't arrived so I think we're done for today. It was good to have you sitting in on the clinic and asking so many questions. I hope you enjoy the rest of your time with us in Neurology.'

I started to gather my notes together, but before Amina could respond, there was a knock at the door.

'I'm very sorry, doctor, but your five o'clock has just turned up. His train had been held at a stop signal and he's only now got here,' explained our clinic nurse.

'OK, that's a shame he's so late, but I can see him,' I said with an air of resignation.

'The trouble is we have to lock up at six. Those are the rules. There isn't enough nursing cover after then. Shall we make another appointment for him?'

As I considered the options, a very elegantly dressed man appeared at the nurse's shoulder, the sapphire flash of his eyes turning to gaze at us intently.

'I'm terribly sorry, doctor, but I couldn't help overhearing the conversation. Michael Buckley's the name and if you can give me even twenty minutes of your time, I would be so grateful. I do understand that I'm late, but it was out of my control, I'm afraid.'

He was striking, verging on the flamboyant, but somehow self-aware enough not to appear contrived. His shock of silver hair was impressively swept back, with the straightest of side partings dividing it with precision. Immaculately dressed, with a spotted burgundy handkerchief in the top pocket of his green tweed jacket, and a navy blue tie knotted in an extravagant Windsor, Michael Buckley was nothing if not majestic. But it was when he spoke that the drama became extraordinary. Out of his mouth came a delicious, sonorous voice that gripped the listener. It was authoritative, yet generous; calm but full of excitement; passionate and at the same time quite gentle. The entire presentation seemed paradoxical yet polished to a high lustre. He made his mark.

'Do come in, Mr Buckley, but I'm afraid it can't be for long, as you heard. This is Amina, one of our students. I hope you don't mind her being with us?'

'Not at all. Thank you so much,' he said with a nod of appreciation.

'What exactly is the problem?' I asked.

Despite his outward demeanour, things were apparently not perfect for Michael. For this singularly impressive man, now in his late sixties, had come to seek help. He was encountering difficulties and, of all things, given his bearing and gravitas, he was worried about his speech.

'I'm having trouble finding the right words. I can be midway through a sentence and then groping for the correct word, if you see what I mean. It's so frustrating and, perhaps more to the point, it's bloody embarrassing. You see, doctor, I pride myself on my ability to speak to the point, be sharp in my responses. Now all of that is fading away. I feel like I'm losing my faculties. It bothers me – a lot.'

He said all this without hesitation but with a clear, genuine fear. If there was a problem with his speech, though, it wasn't easy to detect on first listening.

'I see. How long have you noticed there's been a problem?' I asked, wanting to hear more.

'I suppose it must be for at least a year. It's getting worse.'

'Would anyone notice?'

'Well, I notice!' Michael's voice rose indignantly. 'Nobody has mentioned anything, but I suspect people who know me notice. It's made me lose confidence, you see; I tend not to pursue a line of argument, and just drop it instead. I'd never do that in the past.'

I still couldn't detect any errors in his speech. It seemed flawless. I could see by Amina's face that she thought the same too.

'Tell me a little about yourself,' I asked, hoping to hear an example in his speech of what he was worried about. 'What did you do for a living?'

'I worked in the City – investments. I miss it sometimes but to be honest I'm glad to be free of that life. It wore me down. Now I potter around the garden, take care of the grandchildren, enjoy watching the cricket or rugby.'

His hands interlocked with steepled fingers, Michael went on to tell me how, while he'd been very successful at his job, he thought he hadn't achieved anything very meaningful in life; how his biggest regret was declining an offer to do a PhD in history and how he missed being challenged. It was too late now, of course, but he still loved reading. Had I seen the latest book on the Glorious Revolution of 1688 when James II had been deposed by William of Orange and his Dutch fleet? It was clearly hogwash to suggest that this had been anything other than the most successful invasion of the British Isles. Somehow, history had subsequently been rewritten, distorted to conform to a British narrative of not being conquered.

He was compelling to listen to, and not once did he pause or stumble over his words. I began to think there wasn't going to be much substance to his concerns because he was, literally, so well spoken. But then we turned to the subject of sport.

'Yes, I used to play a lot of rugby. I got to county level. Still love watching it!' He smiled.

'Which position did you play?'

He was quiet, seemingly puzzled and definitely taking far too long to answer a very straightforward question.

'Position?' Michael asked quizzically, raising his shoulders and looking to Amina for help.

'Yes, position,' I said. 'Were you in the scrum?'

'Scrum?' There was another long pause. 'What's a scrum?'

If you're unfamiliar with the game of rugby, a 'scrum' might not mean much to you either, but ask a British schoolchild who has ever seen a game, even on the television, and they would understand what you meant. The scrum is a fundamental part of the game of rugby. It's a peculiarly bizarre means of restarting play. Eight men on each side put their heads down and push against each other, but only when the egg-shaped rugby ball is slipped in between them! To the outsider, it looks like a grotesque,

55

primeval form of tribal warfare. But for a man who played rugby at a high level in his youth not to know the meaning of 'scrum' was very odd. It was a clue.

'You're not familiar with a scrum?' I asked, just to make sure.

'No, I don't think so. What is it?'

'A scrum is when two sets of players put their heads down, come together and then shove when the ball is put in between them.'

'Oh, a scrum, yes of course!' Michael flinched.

'So which position did you play?' I asked again.

'Ah yes, I was a scrum half. I had my face rubbed in the pitch a lot!' He laughed.

I smiled too. I knew exactly what he meant. That sentence brought back a reel of memories for me. What flickered into my mind was the sight and then the smell of sodden mud, and my own face being rubbed in it. As a boy I had found myself in a very traditional English grammar school in Birmingham, an establishment which prided itself on its rugby prowess. Being a tall eleven-year-old meant there was no escape from playing the game, particularly when our very enthusiastic and unrelenting Cumbrian rugby coach spotted you as a potential talent.

My thoughts returned to Michael. Was this apparent aberration of not knowing what a scrum is a mere quirk? That seemed highly unlikely. I should pursue this more.

'Tell me, what do we call something we use to cut grass in the garden?' I asked, my mind obviously still not too far away from soil.

'Um, well, a grass cutter I suppose!' Michael said with a laugh.

'There's another name for it.'

Michael looked anxious. 'I'm not sure. Can't think of it!'

'What about the thing I might use to sweep leaves up in the garden?'

He frowned and looked up at the ceiling as if there might be inspiration there.

'I don't know.' He paused. 'This is precisely the kind of problem I might have, doctor,' he said, and I could see the fear returning to his face.

'I see. I hope you won't mind me asking you a few other questions. This is the only way I get to find out exactly what your problem is.'

He nodded.

'Tell me, do you know what a lawnmower is?' I asked.

Michael snorted impatiently. 'Of course I know what a lawnmower is. I bought one for my youngest daughter the other day. She insisted she needed it and I have to say I was a bit annoyed. It came out of the blue and I don't know why she had to ask me of all people. She's old enough to buy a lawnmower herself, if you ask me!'

Michael was becoming a little agitated, but he still hadn't answered my question.

'So, what is a lawnmower for?' I repeated.

'Well, a lawnmower is for you know . . . what my daughter wants.'

'Does your daughter have a garden?'

'No, she lives on the second floor of a block of flats.'

'So, why does she need a lawnmower?'

'That's a silly question. Of course, everybody needs a lawnmower! She wants to hang it up on her wall.'

'I see.'

Amina's eyes had become wide open.

'Frankly, I'm fed up of her asking for favours!' he continued, raising his hands in apparent exasperation and turning pointedly to Amina, as if she were for some reason a representative of wayward daughters.

An important skill that we try to teach medical students is

not to be deflected from questions if you don't get an answer, unless of course it's distressing to the patient. It would have been easy for me to drop my rather persistent line of enquiry about grass-cutting machines, but continuing with it had revealed something unexpected. Michael really didn't seem to know what a lawnmower was for or what it was.

'What about a rake? What would you use a rake for?'

'A rake?'

'Yes, a rake.'

'Well, that's easy. You use it to . . . you know . . . and then you're done.'

'You use it to . . .?' I left the question hanging.

'Exactly! You use it and then you're done,' he replied confidently.

'Can you draw a rake for me?'

'I'm not good at drawing.'

'Just a rough picture so I can see what you mean,' I persisted.

I gave him my ballpoint pen and he drew, with great skill and in some detail, a toothbrush.

'Thank you,' I said.

Amina, looking over my shoulder, opened her mouth in astonishment. A toothbrush is obviously not a rake but, in some ways, it is related. You don't use it to clear the garden of leaves, but it is a brushing implement. Was this another clue?

There was a knock at the door.

'I'm really very sorry, doctor, but we've cleared all the other clinic rooms and have to shut the entrance now,' explained the nurse.

'Mr Buckley, I apologise that we can't continue today but we can bring you back next week to complete the rest of the testing, and now I have some idea of your symptoms.'

He left, seemingly very grateful for having had the opportunity to see me even briefly.

'Wow! I would never have guessed that he would have any problems with his language at all. He seemed so . . . sophisticated, so fluent,' Amina commented.

'I agree. It really was difficult to spot any problems with his language initially, but do you see how persisting with carefully taking the history can pay off? Wasn't it interesting how, as we went on with the conversation, we began to find out what his difficulties are?'

'Oh yeah.' Amina smiled. 'Even I know what a scrum is.'

~ o ~

Dusk was falling over Bloomsbury, draping the streets with long shadows as I made my way to the tube station at Holborn. The street lights had not yet been switched on. My thoughts went back to the symptoms that Michael was experiencing. Why was he having difficulty with his language? To my disappointment there was a large crowd gathered outside the entrance of Holborn station which had been closed, supposedly because the number of people inside was dangerously high. In typical British style, there was no noise, apart from the man who was handing out free newspapers shouting out a headline about the latest spate of cycling fatalities on London roads. The commuters stood still as they waited patiently in the gloaming. I decided that this was not for me and walked towards Piccadilly Circus.

On Shaftesbury Avenue, theatregoers were milling outside some of the big show venues. I cut into Chinatown and then down Gerrard Street. As I passed the New Loon Moon Chinese supermarket, my eyes happened to glance up to a bottle-green plaque on its facade. In clear white, startling text it stated: 'Here in the former Turk's Head Tavern, Dr Samuel Johnson & Joshua Reynolds founded The Club in 1764.' The irony made me smile. That afternoon I had seen a man who was having difficulty remembering words and here was the haunt of the great

59

lexicographer, Samuel Johnson, the man who had produced the first English dictionary of any note.[1]

This venue had apparently been the pub where in the eighteenth century some of the most famous men in Georgian London met on a regular basis. Johnson and Reynolds, artist and founder of the Royal Academy of Arts, led an elite group which included the economist and philosopher Edmund Burke, the actor and playwright David Garrick, Edward Gibbon the historian, Oliver Goldsmith the novelist, the playwright and poet Richard Sheridan, Adam Smith the economist and Johnson's biographer James Boswell. They met every Monday at seven in the evening as the 'Literary Club' in the Turk's Head where apparently any topic might be discussed, except politics.[2]

Johnson, from all accounts, was an extraordinary sight. William Hogarth – painter, social critic and cartoonist – unaware of his true identity, thought that he must be some monstrous idiot when he first met him. Standing nearly six feet tall, Johnson towered over his fellow citizens. His face in parts swollen and scarred by scrofula (tuberculous lymph nodes), his head frequently making sudden thrusts, and his mouth emitting explosive grunts, whistles or clucking noises, Johnson would have been quite an alarming figure.[3] His great friend Reynolds, who was commissioned to paint the portraits of both King George III and Queen Charlotte, also depicted Johnson on a number of occasions – including a remarkable study of the man without his shabby wig but poised with his hands contorted in 'odd gesticulations'.

Neurologists who have read Boswell's and other contemporary descriptions of Johnson's movements have been convinced that he suffered from the tics associated with Tourette's syndrome.[4] Apparently, when King George heard he was visiting the royal library, he dropped what he was doing to meet the famous Dr Johnson. This was ostensibly to urge the great man to write more, but probably out of sheer curiosity and a desire to secure a first-hand sighting of the grotesque genius.

As unusual as he might have looked, Johnson was widely lauded for his dictionary.[1] Published in 1755, it had taken him eight years to compile the two-volume folio edition, which contained the meanings of 42,773 words. In his house in Gough Square, just a twenty-minute walk from the National Hospital for Neurology & Neurosurgery and still open to visitors, Johnson amassed hundreds of books, underlining words in them for his assistants with the aim not only to collate these, but also to give examples of how they were used in English at that time. This focus on usage was a crucial part of his thinking.

In the first edition of his dictionary, Johnson included 114,000 quotations to illustrate the usage of words. He understood that definitions change. His solution to enabling readers to appreciate the meaning of a word was to give them examples from literature. These 'illustrations' were an attempt to provide exemplars of how an unfamiliar word might be used: its nuances and the concepts associated with it. His dictionary turned out to be a major success, having a huge influence on how subsequent ones were produced. This included the Oxford English Dictionary, the second edition of which contained over 230,000 'main entries', spanning twenty volumes. Just as with Johnson, there were quotations to illustrate the meanings of words. The longest entry was for the verb 'set', which required 60,000 words to describe some 580 senses in which it might be used.

But how are the meanings of words and the concepts associated with them represented in our brains? Do we possess and refer to a kind of dictionary – a mental lexicon? The vocabulary of the average adult, native English speaker apparently contains approximately ten thousand names for things – nouns – alone. There are words for plants, animals, tools, musical instruments, for different types of furniture, modes of travel, all sorts of things. But why on earth was Michael having trouble finding words?

~ o ~

I had not been able to complete my assessment when I had first seen Michael because there had been so little time. To begin with, I hadn't really determined what kind of language impairment he had. Nor had I established whether his difficulties were confined to language. His trouble with finding the right word or even remembering certain words might have simply been part of a more general problem with his memory. I hadn't tested that. When he returned the following week, my aim was to focus on attempting to answer these questions.

As he entered, Michael smiled slightly nervously. He had brought with him his wife, Sarah. As before, he was very elegantly dressed, this time in a blue blazer, a tie with alternating yellow and red stripes, and eye-catching white chinos, almost as if he'd stepped off a yacht and into the clinic, I thought. Sarah was altogether less conspicuous, wearing a plain brown cardigan and grey herringbone trousers. She held his hand tightly as they walked in.

I started by testing Michael's ability to understand language but found quite quickly that he could follow complex sets of instructions with ease. He was also able to repeat words and long sentences without difficulty and his speech was fluent throughout, without any discernible errors. When I pointed to a series of objects around the room, he was able to name them without hesitation. His difficulties became evident, however, when I presented him with a set of line drawings of animals or objects, particularly when they were uncommon things.

'And what's this?' I asked, pointing to a drawing of a penguin. 'That's a bird,' he said.

'Yes, but what type of bird?'

'Um. Now that's a question!' he retorted.

'Where does it live, this bird?' I persisted.

'God knows! The zoo?' He beamed, rather pleased with himself for coming up with a clever answer.

'Yes, but where did it come from? Somewhere hot or somewhere cold?'

'Not sure, to be honest with you. I'd guess somewhere cold, but I've never been good with birds.'

'What about this?' I asked, pointing now to a picture of a harp.

'Is it a kind of window?'

'No, it's a kind of musical instrument,' I explained. 'Does that help?'

'Not really, but then I was never very good at music.'

'It's a harp. Do you know what kind of sound it makes?'

Michael shrugged. 'Like a piano?'

So it went on. Although Michael could name common objects, like a pencil or a spoon, he really struggled with more unusual items. With animals, he might know that it was a bird but he couldn't tell me which type. When he was shown a picture of a tiger he called it a cat, and he came to the same conclusion when shown a cheetah or a kitten.

Even more telling, when Michael was asked to name as many animals as he could in a minute, he could think of only four, resorting to common examples: dog, cat, rat and cow. And if I asked him to give me the names of different types of dog, he was stuck on one – a labrador. He just couldn't come up with any other examples. Michael was running out of words. His vocabulary seemed to have shrunk, but why? Was it possible that this was simply because his memory in general was beginning to be eroded?

There was only one way to find out. I gave Michael a few words to remember and discovered that he could recall these very well over a period of a few seconds. If I asked him to repeat sentences of increasing complexity, he could also do this without difficulty. His short-term memory, by which psychologists mean memory for information retained over seconds, therefore appeared intact.

Next, I tested his memory over longer periods of time. I gave him a list of words to learn, which he was able to do quite quickly. Then I asked him to remember them because I was going to ask him about them later. I gave him a complex line drawing to copy and asked him to remember this too because I was going to ask him to draw it from memory. Fifteen minutes later he could recall the words well, and also made a pretty fair copy of the drawing that I had shown him. His memory for both verbal material (the written words he had learned) and visual material (the complex figure he had drawn) seemed to be excellent.

'Do you listen to the news on the radio or follow it on the television?' I asked.

'Yes, I do.'

'Anything caught your eye in the last few days?'

'Of course, there's that awful bombing in Syria. It's difficult to imagine how the people are coping there. Their own army has turned on them.'

Michael was referring to the massive artillery barrage that had been unleashed on parts of the ancient city of Homs, a focal point of the Syrian insurrection, which was now bearing the brunt of the conflict, nearly a year on from the original Arab Spring protests. The people trapped there were caught under a rain of mortars and shells, a foretaste of what was to occur in so many other parts of the country.

'Then there are the economic problems in Greece. The people are fed up with austerity and are blaming it all on the Germans!'

With great clarity, Michael went on to recount several other stories that were in the news at that time. I asked him about recent events in his own life and, again, he could tell me in detail about trips he'd made to see his family, a holiday he'd been on in the Algarve and about a play he'd seen on London's South Bank. Sarah confirmed that the details he had given me were correct. All this suggested that his recall of events or episodes in his life was actually quite intact. However, although these aspects of his memory seemed fine, he was struggling with his memory for words and the meaning of words. His problems seemed to lie with his verbal lexicon, his mental dictionary of words and their meanings.

'Would it be possible to speak to Sarah alone?' I asked. 'It's often easier to get a better understanding of how things are at home.'

'Of course. We have no secrets,' Michael said, smiling as he left the consulting room.

'Have you noticed any difficulties that Michael might be experiencing?'

Sarah flicked her greying hair back off her gaunt face and cleared her throat. 'I began to wonder if there was a problem a little while ago, doctor,' she said reflectively. 'Our grandson, Ben, is always cracking jokes and Michael has always loved to hear them, but I noticed two years ago that he wasn't understanding them. He would smile, but I could tell that he didn't get the humour.' Part of the attraction of knowing Michael had apparently been his playful temperament. 'He always made me laugh,' she said. 'He used to have a very mischievous sense of irony. That's all faded away. The fun has gone.'

'I see, so he seems to have lost his sense of humour?'

'I think it's more than that. I could see that he was struggling with the names of some of those drawings of animals and objects you showed him. Sometimes, in conversation, he seems to forget the name of something. He'll say, "Sarah can you pass me the . . .

the . . . the . . . thingy, please?" It could be the butter dish or his spectacles, anything really. He seems to have forgotten the names of things. It's the same if I want him to get something for me. The other day, I was busy on the phone and asked him to bring me a pen to jot down some details. He came back with an envelope. When we were cooking yesterday, I asked him to pass me some garlic. He fished out some parsley. It's things like that.'

'How he is with other people?'

'It's been difficult. His friends don't stay as long when they visit,' Sarah responded. 'They always used to have a laugh with him, but nowadays that's rare, and there are more awkward, stilted conversations with long pauses. If I'm honest, they don't really come to see him anymore. He has to go to them if he wants to meet up, but he's less inclined because he lacks confidence. I know he doesn't feel the same person anymore. He's worried about how his friends view him.'

Sarah hadn't noticed much else. Michael's personality, apart from the loss of his sense of humour, was much the same. He wasn't behaving oddly. In every way, he remained the perfect gentleman as far as she was concerned. He was never rude. He could be relied upon to pick up the grandchildren on time. He drove with care. True, he was using the computer less and now she was doing much of the online banking, but that was about it. And no, he wasn't doing anything odd like hoarding, gambling or drinking excessively as far as she knew.

I brought Michael back in.

'Haven't been spilling the beans, have you, Sarah?' he said with a wink.

'No more than usual, Michael,' she responded, clearly pleased that he was cracking a joke.

'I think I'm beginning to understand the kind of difficulty you're having,' I explained. 'We're going to need to do some more tests to look into this in detail.'

Michael was obviously relieved. 'Thank goodness! I was worried that you wouldn't believe I had any kind of problem. Thank you for taking this seriously, doctor.'

'I'm going to arrange an MRI scan of your brain and some more testing of your language, memory and thinking with one of our neuropsychologists. They can probe a little deeper into your difficulties and perhaps help us to understand a bit more about why you're having problems finding the right words.'

'Oh good! Thank you. I really want to find out what's going on.'

So, to be honest, did I.

~ o ~

Why had Michael's vocabulary shrunk? One thing that most of us might agree upon is that, even if we had the inclination, we don't have the capacity to learn all the entries in a small dictionary, let alone the Oxford English Dictionary. We don't retain verbatim definitions of words, or the many nuanced ways in which they might be used, or examples of how they have been deployed in literature. That's why Dr Johnson put together his massive work in the first place: to save us trying to retain all that information. If the human brain were to store so much information in the way Johnson had designed his dictionary, it would soon become overwhelmed. We nevertheless hold a remarkable amount of information about the world around us: people, places, living things, non-living things, historical and scientific facts, abstract ideas.

How do we do it? Research performed by psychologists in the 1960s first suggested that we do indeed possess a form of mental lexicon linked to a knowledge database of sorts, but this representation in our brains is very different from even a short dictionary.[5] Rather than having a separate entry for each word with definitions for every one of them, our repository of knowledge seems to be set up to understand the relationships between words and, most importantly, the concepts that go along with them. For example, we know that canaries, larks and starlings are all birds. It might therefore make sense in any representation of knowledge to keep all the information that is common to birds close together while things that aren't birds (like fish or cats, or tables and chairs) should be kept further away because they belong to different groups of objects.

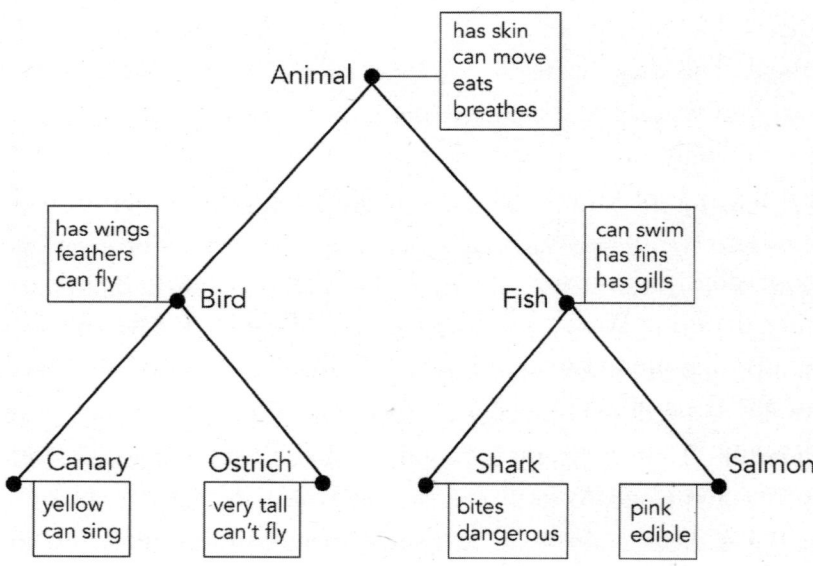

Fig. 2 | Hierarchical classification of knowledge. A semantic network proposal for how information about animals might be stored in the brain. Adapted from Collins & Quillian (1969).[5]

Objects that we classify as being in the same category, some psychologists argued, are closer to each other in the way our brains organise this information. Furthermore, they are stored in a hierarchical system, like a branching tree of knowledge (**Fig. 2**). Concepts act as rules for the way we categorise information. A particular super-ordinate category, like bird, would include all the examples of the concept bird. This format appears to be an economical, compact way to retain information about items that are similar because they have close conceptual links between them. Instead of keeping separate definitions of canaries, larks, starlings and any other birds, we can keep a single entry that brings together the attributes that are common to all of them, and indeed most other birds. In other words, we have a generic definition for birds. This might, for example, specify that in general they all have beaks, feathers and the ability to fly.

This kind of insight, I thought, might also be important in what I had observed in Michael. He could name only four animals, and these were all common exemplars: dog, cat, rat and cow. And when asked to name different breeds of dog, he could think of only one, a relatively common one – a labrador. It was as if his 'tree of knowledge', the way he represented words, had been severely pruned back to just a few branches. He knew about dogs, but seemed to have lost the names of any breed other than one. Was it possible that he was running out of words because his mental lexicon, not only for the words themselves but for their conceptual meanings, was slowly degrading? Was that why, in our previous meeting, he didn't know what a lawn-mower was or how was it used?

In the 1980s, researchers who studied how our brains store words and their meanings started to question the original proposal that knowledge was represented in a tree-like, hierarchical structure (**Fig. 2**). One particular line of work that led to this questioning was a set of studies into how we represent some unusual exemplars of a category of items; for example, ostriches and penguins in the superordinate category of birds. Neither ostriches nor penguins fly. So, we'd expect these uncommon birds to have some other attributes attached to them in our representations of these words: one of them swims, the other has very big legs, and so on. Conceptually, though, each one is a bird. But, clearly, they are different to most other birds we know.

Indeed, when psychologists tested how quickly people classified canaries and ostriches as birds, they found that they perform this classification much faster for canaries. Yet the model shown in **Fig. 2**, on face value at least, might predict there would be no difference in speed of response because ostriches and canaries are equidistant from the superordinate category of bird.

Findings such as this led scientists to question the hierarchical system that was originally proposed to explain our database of semantic knowledge. Instead of agreeing with this view, they developed new models of how words and their associated concepts are stored in the brain.[6] The inspiration for the structure of these models came from three key insights that neuroscientists had gained about how neurons (nerve cells) in the brain work. The first was that neurons are connected to many other neurons, allowing for parallel distributed processing (PDP), so that signals can be broadcast simultaneously across many nerve cells. The second was that the strength of these connections between neurons can alter with learning: some connections can grow stronger if they are frequently active, whereas others can become weaker. Finally, networks of neurons can receive signals when errors are made, so that they can effectively learn from their mistakes. In fact, the error signals can propagate through a network to strengthen or weaken some of the connections between neurons that might have led to the mistake.

These new models are sometimes referred to as PDP 'connectionist models' because they consist of networks which have connections between 'units' representing neurons in the brain. They represent concepts in a completely different way to conventional hierarchical systems. In connectionist models, a concept such as a canary is represented by the pattern of activity across the entire network, not just by a node within a hierarchy. Initially, a neural network may not be very good at differentiating a canary from other living things, let alone from ostriches. However, with training through many iterations of learning, where the network is fed back its mistakes (by a process known as 'back propagation'), the connections between the units in the network slowly alter. Some get stronger, others grow weaker until the network can perform extremely well at classifying a concept, such as a canary, and all the properties that belong to it. Not only that, but it can generalise to other related concepts. Because similar concepts will have very similar patterns of activity across the network, a new concept, such as say a starling, will be learned very quickly.

These connectionist approaches have been found to be highly effective in accounting for several aspects of human semantics. They have, for example, shown how it is possible for the distributed pattern of activity within a neural network to learn concepts just as children do during development, and crucially also how this knowledge might be degraded in brain diseases.[6] These approaches to semantics were the forerunners of the artificial intelligence revolution, which has drawn on neural networks and back propagation as key mechanisms with which to solve a wide range of problems. Could the organisation of semantic knowledge also be relevant to understanding the kinds of problem that Michael was experiencing?

~ o ~

A couple of weeks later, Michael went on to receive detailed cognitive testing from one of my neuropsychology colleagues. She agreed that his longer-term or episodic memory was intact. When we speak of memory, most of us usually mean recollections of events – episodes from our past. This is what neuropsychologists refer to as episodic memory, and just as I had found on brief testing, she also concluded that this particular cognitive function was working well in Michael. However, like me, she observed that he encountered great difficulty in naming pictures and even some real objects that she showed him. This could be whether he was seeing, hearing or touching an item. For example, he couldn't find the name for a set of keys when he was shown them, but he also failed to recognise them when he simply heard the keys jangling, or when he was allowed to touch them with his hand with his eyes shut.

Michael's difficulties, she noted in her report, extended to matching words to pictures. When he was shown a word such as 'swing' or 'watch', he couldn't pick the right object that matched these words from a selection of pictures. So, he definitely had a problem in finding the names of things, but his difficulties extended beyond naming. She had found that he had great trouble in sorting objects – and this was very telling to me – according to how similar they were. For example, most people consider a stool to be similar to a chair, as is a sofa; whereas a bin and a lamp are very different. Michael apparently thought all these domestic objects were similar, with no difference in the degree of similarity between them.

He was also unable to demonstrate the proper use of many of the objects she showed him. For example, when presented with a screwdriver he used it like a spoon, and when given a hammer he used it like a saw. He didn't seem to know what a key was for at all. It wasn't, therefore, simply that he was suffering from a problem with naming objects. It was understanding the concepts associated with these objects – it was about their meaning. Michael, she concluded, seemed to be losing his knowledge about the world and the things that occupy it – his 'semantic memory'.[7]

Semantic memory is the term psychologists use to refer to our knowledge base. Do you know how many months there are in a year? What's a fish? Is a shark a fish? What about an octopus? How many different ways can you use a knife? Think about it: your answers to this question depends upon the context. You can use a knife to cut bread; but a different action is required to spread butter on toast; and still a different technique is required if you want to scoop some jam out of jar with a knife; and, of course, there's yet another completely different way of using a knife if you want to harm someone. Somehow all this information accompanies the word 'knife'. We carry it with us as we've learned about the uses of a knife. We even know that there are different types of knives: ones for cheese, or to cut steak, or perhaps peel fruit, or cut bread. This type of knowledge is referred to as semantic memory; it is our conceptual understanding. It comes to cover a vast territory as we learn about the world we occupy.

Thus, while Michael's episodic memory was intact, he seemed to be losing his semantic memory. That's what sometimes made it difficult for him to find the right word in conversation. This tallied with the impression that I'd obtained in the clinic. His difficulties in conversation went deeper than being able to retrieve the correct word, though. He was struggling to understand what that word meant, the concepts associated with a label such as 'scrum' or 'lawnmower'. Our understanding of a word is not just tied to a tag or a label – the word itself – but it is also intimately connected to the concepts that are linked to that word.

The Colombian author Gabriel García Márquez appreciated this well.[8] In his landmark novel, *One Hundred Years of Solitude*, the inhabitants of his fictitious town of Macondo suffer from a plague of insomnia. But it is a very unusual form of insomnia, one which leads to the 'loss of the name and notion of things'. One of the townsfolk becomes so aware of this that he takes to sticking labels on objects (table, chair, clock, door and so on) simply so that he can remember what they are called. However, 'he realised that the day might come when things would be recognised by their inscriptions but that no one would remember their use'. In other words, even if they had the label – the word

73

for an object – the townsfolk wouldn't know the concept that goes with it: how to use a table or a chair, how to milk a cow, or what to do with any of the objects or animals around them.

Concepts aren't just linked to objects. They're also associated with abstract words, like truth, love, democracy. And by combining concepts we have the ability to create a huge, almost infinite, combination of ideas and thoughts. These can be novel or highly creative thoughts, but they're all built on simpler ideas which we construct effortlessly with the words that we attach to them. The label is simply shorthand for a rich set of information – knowledge about the concept stored with the label.

Consider, for example, the word 'English'. It means so much more than the English language. It captures a vast range of qualities and characteristics, some historical, some social, many about the sort of person who we might anticipate to be English. Even without all this knowledge, as a brown child walking down the street in 1960s London, I came to know very quickly that I definitely wasn't 'English'. The names people called me made me understand that concept quite vividly. I recall that, as an excited nine-year-old, when I first moved with my family from London to live in Birmingham, I was at first very disappointed. I had great difficulty making sense of what the locals – the 'Brummies' – were saying. Their accent sounded completely foreign to me. 'Lovely people, but I don't understand them,' I wrote emphatically in my diary.

Soon, though, I began to realise that there were different types of 'English' people: urban, rural, northern, Brummie, cockney and a whole lot more, all with different ways of speaking. In fact, there are many different types of northern accent alone: from Yorkshire or Lancashire, from Liverpool or Manchester, from Newcastle or Cumbria, and so on. The nuances between the concepts attached to each of these characteristics of Englishness are vast, but they only came into my semantic knowledge base with experience accumulated over many years. So, I understood that semantic memory grows with time, but could it also become degraded?

~ o ~

Michael's scan came back a few days later. I received an alert to tell me it had been reported. When I looked at it, the result was obvious. Much of his brain looked pretty normal, but the left temporal lobe was definitely abnormal. In fact, the very tip of it – the temporal 'pole' – was shrunken and withered.[9] It was 'atrophied'. This is the characteristic finding in a relatively rare condition called semantic dementia, in which people progressively lose their semantic memory (**Fig. 3**).[10] In fact, rather astonishingly, this loss specifically of semantic memory is exactly what García Márquez described as occurring in his fictional account of what happened to the people of Macondo.

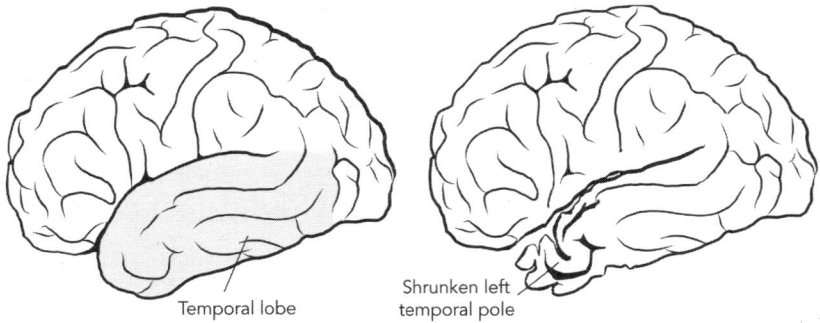

Temporal lobe

Shrunken left temporal pole

Fig. 3 | **The brain in semantic dementia.** On the left is a normal healthy brain with the left temporal lobe shaded. In semantic dementia, the very anterior part of the left temporal lobe, called the temporal pole, shrinks (as shown on the right-hand image), leading to loss of semantic memory.

Regardless of precisely *how* semantic knowledge is represented in our brain (whether it is a hierarchical classification system or more like a neural network, connectionist model), studies of healthy people as well as patients with semantic dementia have now led to the consensus that a key part of that representation is located at the very tip of the left temporal lobe.[11] This semantic store is multimodal: it holds the look, sound, touch, even the smell and taste of an item, if these are relevant. So, when a patient with semantic dementia begins to lose their understanding of the word 'keys', they also start to encounter difficulty in accessing all the attributes associated with keys – what they are made of, their size and shape, their weight and feel, the jangling sound they make when a set of them is shaken – as well as what keys are and, most importantly, how they are used.

Intriguingly, detailed testing of people with neurological conditions, including semantic dementia, has revealed that the representations of living things might be stored separately to non-living objects within the brain.[12] Some patients can show impaired knowledge of living things (especially animals, fruit and vegetables) while retaining a relatively intact understanding of non-living things (objects such as tools), while others show the opposite. These findings offer further evidence to constrain how scientists model the architecture of semantic knowledge representations in the human brain.[13]

When I saw them next, Michael and Sarah were keen to find out the results of his scan.

'I'm afraid it isn't normal,' I said.

'No? Not normal. In what way?'

'One part of the brain is smaller than it should be. It has shrunk.'

'Oh!'

Sarah gripped her husband's hand.

'Yes, and I'm afraid it's that part of the brain that is involved in holding a store of the meanings of words. The results of the scan fit with your concerns about not being able to find the right word at times.'

Michael seemed partly relieved that something had been found to explain his symptoms, but he was also clearly worried. 'I see. Which part of the brain is it?'

'Let me show you.' I turned my computer screen so he could see the scan as I pointed to the abnormal part. 'You see this left part of what we call the temporal lobe? It's much smaller than on the right side of the brain.'

'And does it matter that it is on the left?' he asked.

'Yes. For most of us, language functions are in the left hemisphere. There are lots of different components of language. The processes that give us our understanding of words and their meanings are thought to be located here, right here in the tip of the left temporal lobe.'[14]

'And that's why I'm having problems?'

'I'm afraid so.'

Michael closed his eyes. When he opened them, he avoided my gaze and asked what everyone wants to know.

'Is this going to get worse?'

'I think that is likely – but it may occur very slowly.'

'So eventually I'll run out of words?'

I didn't respond. It wasn't just words that Michael would struggle to remember but also his knowledge of the world – the concepts that go with the words. It was difficult to say this to him. I had just given him the diagnosis and it would be miserable to say all this at the same time.

'Everyone's different. Some people progress very slowly.'

~ o ~

Semantic dementia can be hard for a doctor to spot. Most won't even be aware that it exists. Those who do may not be able to find anything remarkable in a patient's speech unless they assess it carefully and consider the patient's history with that specific diagnosis in mind. Family members and friends are sometimes invaluable in picking up important clues. Often patients' speech reveals a tendency to use generic terms (such as 'animal') instead of a specific example (like 'dog'). As Sarah had noted, Michael had been using the term 'thingy' when he couldn't think of a word. Even in their drawings, people with semantic dementia may start to generalise, revealing their lack of knowledge of specific details. A patient asked to depict different animals might resort to drawing a 'default' form with very few of those detailed characteristics that distinguish a fish from a bird or a dog (**Fig. 4**).

Sarah's observations about Michael losing his sense of humour were also quite insightful. To understand a joke requires a person not only to have the vocabulary to comprehend what is said, but also the elaborate concepts linked to the words that are uttered. Mark Twain, a man celebrated for his turn of phrase, once described an acquaintance as 'a good man in the worst sense of the word'.[15] Taken at face value, this comment might be perplexing because 'good' doesn't seem to be used in a familiar way here. The humour emerges only if our conceptual understanding extends beyond the plain meaning of 'good'. In the way Twain paints him, this man is possibly self-righteous, perhaps hypocritical – certainly not at all 'good'. The negative meaning that is conveyed is the result of Twain's audience having an understanding more subtle and varied than a simple definition of the word.

The semantic theory of humour developed by the linguist Victor Raskin proposes that the punchline of a joke works only if it leads to the audience abruptly shifting from the more obvious, primary meaning of a word to a secondary, often opposing, way in which it is being used in the joke.[16] But to understand that shift requires a richer semantic understanding of the information associated with a word. Ironically, in the case of Mark Twain's acerbic comment, 'good' can also be used to mean 'bad'.

For Michael, that sophisticated conceptual understanding of meanings beyond the surface definition of words was seemingly

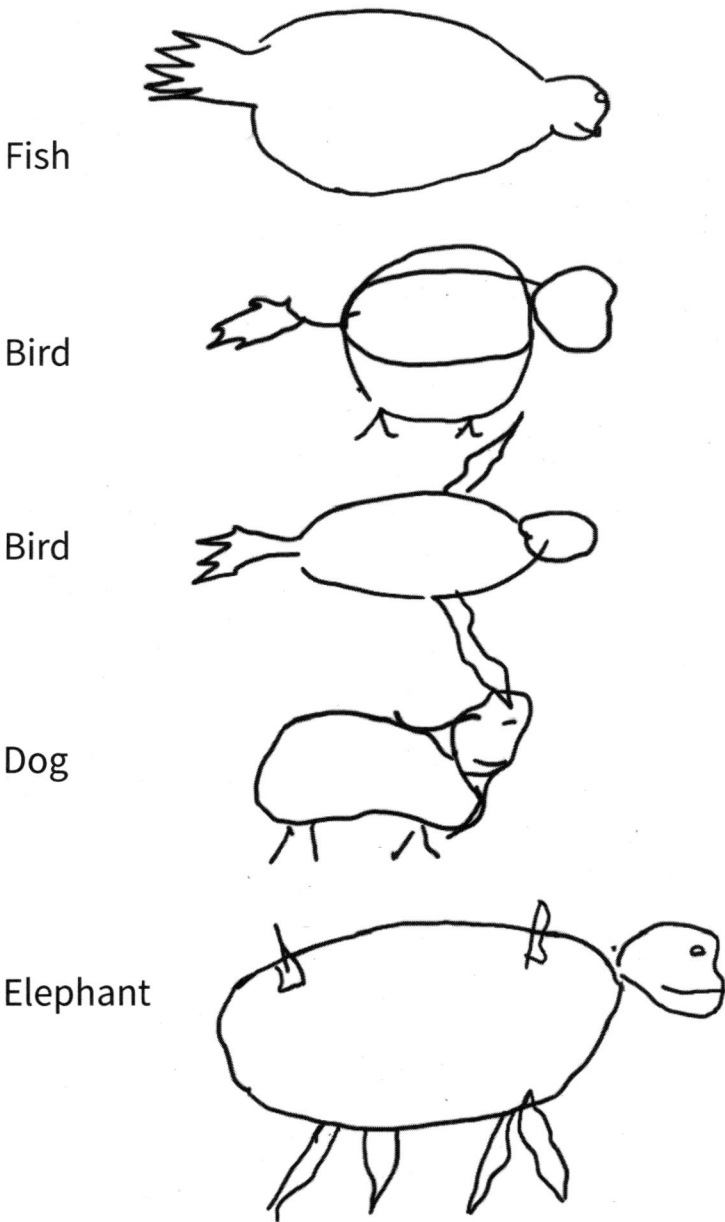

Fish

Bird

Bird

Dog

Elephant

Fig. 4 | Drawings by a patient with semantic dementia. Note how each animal has a very similar generic form, with the fish lacking fins, the top bird having no wings and the elephant missing a trunk. Adapted from Rascovsky et al. (2009).[8]

the first thing to go. Without an understanding of their extended meanings – not simply their dictionary definitions, but all the other information that is associated with them – it had become difficult for him to comprehend why something was funny. His interactions with friends had slowly become more limited, far narrower in scope, but this wasn't because of his conversational ability. He could clearly talk, sometimes in a very compelling fashion, as I had observed when I first met him.

But retaining connections with people involves more than the straightforward exchange of information. We also need to enjoy their company, one key element of which is the mutual appreciation of what we find to be funny. Laughter is crucial for social bonding and may even be an important evolutionary mechanism to influence our sense of belonging to a group.[17] Shared humour motivates us to stay in touch. However, such humour requires a common understanding of the wider semantics associated with the words we use. If we lose this, it can become difficult to hold on to friendships, as Michael was evidently now finding.

For me, as an immigrant child growing up in Birmingham, even after I had become accustomed to the accent and familiarised myself with the nuances of the local slang, I still found British humour bewildering. Maybe I shouldn't have been too worried, as I now know that it has a reputation around the world for being baffling. Sometimes considered rude or, worse, offensive, it can be hard for an outsider to appreciate why sarcasm or irony, two common features of the British sense of humour, are considered to be so very funny.

For example, in the 1970s television series *Fawlty Towers*, an American guest asks Basil Fawlty, the proprietor of the eponymous hotel, 'Is there anywhere they do French food?' Fawlty responds, 'Yes, France, I believe. They seem to like it there. And the swim would certainly sharpen your appetite.' Taken at face value, this might seem at best bizarre, at worst extremely rude.

But in Britain it is deemed extraordinarily funny in the context of the character of Basil Fawlty and his misanthropic, snobbish approach to his chosen profession.

Appreciating humour is now recognised as an important example of how semantic knowledge is crucial for a person's social identity. It is important not only as a way to retain membership of a social group but also to gain entry into one. To make effective, lasting connections – to become and remain part of a group – requires an understanding of how the members of that group conceptualise the world, including what they find funny. Without such an understanding, any relationship is likely to be superficial. This was the predicament Michael now found himself in.

~ o ~

Sarah was anxious to find out more about semantic dementia, to know whether anything could be done, or if there were any clinical trials of new treatments. I fielded her questions, one by one. No, Michael's condition was not Alzheimer's disease. The shrinking of the left temporal pole that occurs in semantic dementia is very unlike what happens in Alzheimer's disease. In that condition, a different part of the brain – called the hippocampus – atrophies and with it goes episodic memory, but semantic memory can be left intact. By contrast, in semantic dementia the hippocampus can be preserved, and this was consistent with the fact that Michael's memory for personal events was intact. Most patients with semantic dementia have an abnormal protein (called TDP-43 or transactive response DNA binding protein 43) deposited in their brains,[18] whereas a very different protein is deposited in the brains of patients with Alzheimer's disease. Semantic dementia and Alzheimer's disease are two different diseases.

Unfortunately, there weren't any new drug trials, although some people were advocating intense therapy that involved relearning vocabulary, rather like a child first learns new words by looking at pictures of objects and naming them. And no, there wasn't any evidence that changing one's diet, or anything else for that matter, would affect the trajectory of the illness – and sadly, no, we don't understand why some people got it and others didn't.

Michael was uncharacteristically quiet.

'But how have you been since I last saw you, Michael? Do you have any other questions for me?'

'I've been OK, thanks. I'm just a bit . . . a bit ashamed.'

'Ashamed? Why's that?'

'Well, I'm ashamed I've got this and all the . . . eh . . . eh . . .' He was trying to find the right word. 'Trouble! All the trouble I'm putting everyone to,' he blurted. 'It's embarrassing . . . and I don't really feel as if I fit in . . . belong . . . with everyone else.'

Sarah held out her hand to him again. 'Darling,' she said kindly, 'you'll always belong.'

But unfortunately, that wasn't to be the case.

~ o ~

The clinic waiting room was busy. Several of the neurologists' lists had been overbooked. There were many urgent referrals and we simply did not have sufficient capacity to deal with the level of demand in the National Health Service. People were evidently getting frustrated. It didn't help that it was raining heavily, and everyone had come into the hospital bedraggled and dripping.

'Does the doctor know I'm here?'

'Is it likely to be long?'

The nurses were doing their best to appease our patients by offering cups of lukewarm, milky tea to anyone who looked impatient. This was Britain, after all, I smiled.

In one corner I spotted Michael. He was still dressed very elegantly but he wasn't sitting down. I looked more closely. He was stroking one of the leaves on a rubber plant in the waiting area, with great tenderness. As I approached, I could hear him saying, 'There, there. Don't be frightened. No tea for you.'

Other patients were deliberately looking away, keeping a safe distance from him. I tapped him on the shoulder.

'Aha, the butcher! So good to see you.'

Sarah, who was sitting on a chair by his side, flinched. She was about to correct him, but I extended out my palm to stop her and smiled at Michael.

Michael squeezed my hand tightly. 'So very good!'

'Very good to see you too.'

'And you, too. So very, very good to see you,' he continued.

When we eventually got into the consulting room, Michael was beaming.

'He's extremely pleased to see you,' said Sarah. 'He thinks the world of you because you were able to explain what he had.'

Michael nodded forcefully.

'How have you been?'

'Well, well. Very well.'

'What have you been up to?'

'A lot of things. Rugby.'

'Rugby?'

'Yes, yes, yes. Rugby.'

'He went to see England play at Twickenham over the weekend and really enjoyed it,' explained Sarah.

'Yes, Twickers! Had a great time. Yes, Twickers!' Michael grinned, but there was no more.

It had been three years since the initial diagnosis and clearly Michael's language had changed considerably over the intervening period. His speech was now far more limited, his vocabulary far more contracted. When I spoke to Sarah alone, having guided

Michael back to the waiting room, she confirmed that he was now unable to communicate effectively. He could speak, even repeat words and sentences, but his lack of understanding of what words meant was having a profound impact. He was often unable to comprehend what people were saying.

It wasn't just jokes or humour that he had difficulty with now. At first, he found infrequently used words challenging. 'Embarrassing', which he had once used in his conversation with me, had become a puzzling concept, but now his bewilderment was apparently extending to common words. His regular vocabulary had shrunk, ever smaller, to a set of nouns like 'house', 'car' and 'food'. Sarah had been surprised that he had used 'rugby' when talking to me.

'It's very hard,' she confessed. 'I've lost the man I married.' She frowned, holding back her tears.

Sarah went on to explain that although she and the children had initially been very sympathetic, they had become increasingly exasperated by Michael's inability to understand even simple instructions. In turn, he had grown irritated with their responses towards him, which slowly became nothing more than perfunctory nods or shakes of the head. Hardly any of his friends came to visit him.

He would often go out driving to get away, but now was no longer able to do so. He had lost his driving licence, not because he had any difficulty driving, but because he was no longer able to understand what road signs and symbols meant. On one occasion, Michael had gone down a road with a 'no entry' sign to find himself driving headlong into oncoming traffic on the wrong side of a dual carriageway. He was lucky to be alive.

At home, his behaviour had become similarly bizarre. He would often attempt to flush clothes down the toilet because it seemed that he thought it was a washing machine; he'd urinate in plant pots, mistaking them for toilets; and, worst still, he would try to eat rotting food that had been thrown away in the bin. Petting the rubber plant in the waiting room was nothing, Sarah concluded.

Outside, through the window of my room in the clinic, I could see the rain still pelting down over the Square, beating a disturbing tattoo over the pavements and car roofs, drenching even those people with their umbrellas held aloft. It was grim. Inside, the atmosphere had also become leaden.

'I'm so very sorry to hear how things have deteriorated.'

Sarah shrugged. 'No news on any new treatments I suppose?' she asked.

'I'm afraid not.'

There was a gentle knocking at the door. It opened slowly and there was Michael holding a cup.

'Tea?' he enquired. Smiling, he came over to Sarah. Placing the cup down gently on my desk, he put his arms around her shoulders and held her very tight as he kissed her on the cheeks. One thing was for sure, he clearly understood her pain.

Sarah closed her eyes. 'He's very much still my husband.'

'Thank you,' Michael said, nodding at her. Looking at me with those wonderful sapphire eyes, he explained, pointing to his mouth, 'No words, but thank you.' And with that he put his arm though hers and they walked quietly out of the clinic into the rain.

3

Losing my memory?

She marched into the consulting room with purpose and an air of self-assurance. He looked beaten by his circumstances. Trish was a short, slight woman, almost birdlike in her appearance. She wore a large black padded coat with an elaborate fleece hood, a creamy collar of plumage which almost threatened to over-whelm her. But somehow you knew that she was in charge. Perhaps it was the emphatic cadence of her entry, the stomp of her heels sharply beating a regular meter. Perhaps it was the way her eyes fixed straight upon me, her face narrowing quizzically, assessing with evident disdain the specialist she had come to see. Whatever it was, Trish looked so much more confident than the man trailing behind her.

'Nice to meet you, doctor,' she said with conviction as she shook my hand firmly.

'Good morning,' I responded. 'And you must be?' I asked her companion.

'I'm Steve, her partner,' he explained quietly.

'He wanted me to come here,' she interjected. 'I'm sure you'll say that it was all a big mistake, and that there was absolutely no need, and we'll be done with it. I know that we're wasting your time, doctor, but he always worries, does our Steve,' she added with a grin.

'Does he? Well, perhaps I might start by asking you what he's concerned about.'

'He thinks I'm losing my memory, but there really is nothing to get worked up about. I'm just not as good as I used to be about remembering stuff – but who is, doctor, who is?'

I nodded, noticing the slight Irish lilt to her accent.

'How long do you think there's been a problem?' I asked.

'But there isn't a problem!' she said, beaming broadly with a dramatic shrug of her shoulders. Steve closed his eyes. He'd obviously heard this many times before.

'I see, but what kind of things has Steve become worried about?'

'Well, nothing really. Of course, I forget a few things here and there, make a few mistakes, but we all do. I'm sure you do too. Don't tell me you don't, doctor?' she enquired, fixing me with a smiling stare.

'So, apart from your memory, is there anything else that's a problem for you?' I asked, avoiding her question.

'Not a thing. I'm as bright as a button,' she said, looking very pleased.

I went on to take the rest of the history from her, but this didn't reveal a great deal. Trish was in her mid-fifties and worked as a receptionist for a firm of solicitors. She'd been born to Irish parents who had moved to the UK. You couldn't take the Irish out of her, as she made clear several times during our conversation. Trish had apparently married young and regretted doing so. She had two children from that relationship before she got divorced ten years ago. She'd met Steve shortly afterwards. Her children had now left home and were working. She didn't have any past medical history of note. She didn't take any medications, smoke or drink alcohol excessively. Her family history was equally unremarkable: there was nobody that she knew who had suffered from memory problems.

'Would you mind if I spoke to Steve?' I asked.

'Not at all, doctor. You go right ahead. You're not going to say anything about me that I might object to now, would you, darling?' she asked mischievously as she turned her gaze towards Steve.

'I meant on his own, without you. We often find it helpful to get the perspective of someone who knows you well,' I explained.

'Oh, I see,' she said a little taken aback, her lips opening momentarily, revealing her apprehension. 'Well, of course, yes.'

I took Trish out into the waiting area. When I returned to the consulting room, Steve was leaning forwards, his head down, cupped in his hands. His posture captured his mood.

'I hope you don't mind me talking to you alone?' I asked.

'No, no. I'm really very grateful. I never usually get the chance to explain things about Trish to a doctor without her being there – so, thank you.'

'What has been the main problem from your point of view?'

'Where do I start? Her memory can be appalling. It's un-believable at times and it's driving me mad.'

'When do you think she was last completely normal?'

Steve told me how he had noted that Trish's recall of infor-mation had been deteriorating over four years. It had started in a very mild manner, hardly noticeable at first, but now she was making mistakes at work, forgetting to enter information into her firm's calendar, double-booking appointments and, most disconcertingly, not recognising important clients when they came for meetings. At home, she might forget conversations. She couldn't be relied on to pay credit card bills on time or to keep important appointments. Her personality was apparently unchanged, although she was now behaving oddly at times.

'How do you mean, oddly?' I asked.

There was a long pause. Steve was clearly reticent to explain but eventually he spoke.

'It's got to the point where she doesn't even seem to recognise me sometimes. We went on a little break to Cornwall a few months ago. It was nice to get away. We had a lovely few days at this fishing village – really lovely – but when it came to the time to leave, it became clear she didn't know who I was.'

'How? How did it become clear?' I asked.

'Well, I told her I'd get the car ready and then we'd drive back home to London, but she became very anxious. She was fretting at the prospect. I mean, she was really worried. I didn't know why at first. But do you know what she said?'

I shook my head.

'She told me that I couldn't possibly go home with her because Steve, her partner, would be there and he definitely wouldn't be pleased if she came home with me. Can you believe that! I nearly keeled over. She said that she'd had a fantastic time but she was so very sorry. It would be too much for Steve. He was the jealous type. She promised she'd keep in touch, but she couldn't possibly take me home. She was going to drive back alone. She really couldn't give me a lift.'

'So, what did you do?'

'I had to think on my feet because she was adamant that I wasn't coming with her. She'd already sat in the car with her bags, but none of mine. I told her that I was a mate of Steve's and she could try calling him to check that it would be alright to give me a lift. Somehow, I managed to persuade her to do that, and while she called on her phone from the car, I walked away and responded to her on my mobile. I told her that it would be perfectly fine for her to bring along my friend. He was a good mate and there was nothing to worry about; she could give him a lift. She eventually agreed to drive me home. It was crazy.' He paused.

'Let me get this right,' I said, trying to make sense of this remarkable story. 'Before you mentioned that you were a friend of Steve's, who did she think you were?'

Another long pause. 'She seemed to believe I was her lover.'

'But not that you were a second Steve?'

'Well, it's difficult to know because what she says can vary. Another time, a couple of weeks ago, in our house, she said I'd better leave before Steve came home. I had to tell her I was Steve. When she laughed as if she didn't believe me, I showed her a photo on the mantlepiece of us together. Of course, then she said she was only joking. But I know she wasn't.'

'So,' I agreed, 'that does sound like she might not recognise you.'

'That's what I thought, but I think it's worse than that. One evening last week she said, "I wish all you Steves would sort yourselves out and decide who is staying with me tonight." I think at times she really believes there are several of us. Is that even possible?" he asked, his face a picture of confusion.

I nodded. 'Sometimes that can happen.'

Steve's initial descriptions had made me wonder whether Trish was having difficulty recognising people, but it seemed that at times she thought there were duplicates of Steve.

In 1923, the French psychiatrist Joseph Capgras described a woman, also in her fifties, who had become paranoid. She was convinced that members of her family, including her children and husband, had somehow been supplanted by people who were 'doubles'. The patient, Madame M, was suffering from a delusion in which she believed that people she knew had been replaced by imposters. Ever since then we have referred to this phenomenon as the Capgras syndrome.[1] It seemed that sometimes Trish really was convinced that Steve had his own set of doppelgangers.

Just three years after Capgras's description of his famous case, his colleagues described a woman, also in Paris, who had come to believe that she was being pursued by two famous actresses of that period, Sarah Bernhardt and Robine. In her mind, they were able to take the form of people she knew: they could enter other people's bodies, including those of her friends, and in this guise these actresses tormented her. This paranoid belief – of different people being the same person in disguise – was termed the Fregoli delusion, in honour of the Italian actor Leopold Fregoli, who had become famous around Europe for his speedy transitions between impersonations of different celebrities such as Verdi, Wagner or Rossini. Both the Capgras and Fregoli delusions are now considered to be part of a spectrum of delusional misidentification syndromes. Observed in some people who develop paranoid thinking such as in schizophrenia, they sometimes occur in patients with other brain disorders too.

'She's also repeating herself,' Steve continued. 'I'll answer a question, but she'll only ask me the same thing again a few minutes later, and then a few minutes after that. And then sometimes she just makes stuff up that never happened, but she's convinced it did. It's driving me mad. There must be a reason, doctor.'

I brought Trish back into the consulting room and asked Steve to sit in the waiting area while we chatted.

'I hope he hasn't been saying anything naughty about me?' she asked with a cheeky smile, but she was clearly anxious to learn what had been discussed.

'Oh no, not at all. I think he's become worried about you. Tell me, do you follow the news? What's going on in the world?'

'I do and I don't,' responded Trish. She had obviously developed her defences well.

'What about over the last few weeks. Has anything caught your eye?'

'Well, there's the whole impeachment issue in the States.'

'Impeachment?' I asked curiously.

'Yes, you know. Nixon! Watergate's been devastating for the Yanks. I feel for them. Of course, you would be devastated, if you found out that your president was such a liar.'

Trish went on to explain what had happened in the Watergate scandal but didn't seem to appreciate that she was speaking about events that had occurred decades ago.

'Anything else that's been important in the world?'

'Well, there's the Middle East. There's always a problem there!' She smiled. 'This time it's awful. I saw the terrible news about those children who were attacked with chemicals in Syria. That was terrible.'

She went on to describe in some vivid detail the chemical weapons attack in Ghouta which had indeed happened only recently, now more than two years since the initial violence in Syria had begun. But her memory let her down with her next sentence.

'Nixon has to do something. If he can bomb Cambodia, he can do the same in Syria now, can't he?'

Trish was confabulating, mistaken in her belief that Nixon was still president of the United States. This, I thought, was probably what Steve had meant when he had said that she made things up, while remaining convinced that she was right.

'Do you know who Barack Obama is?' I asked, curious to know what she would say.

'Of course, he's trying to beat Nixon in the next election. Lovely chap. I'd vote for him.'

Later, I found out from Steve that the night before they'd been watching a television documentary about the Vietnam war and Nixon. Somehow those events that took part decades ago had been transposed and interwoven with those of the present.

'Tell me, Trish, have you been away for a holiday or a break this year?'

'Yes, Cornwall.'

'How was that?'

'Absolutely stunning. I had a fantastic time. I'd go back in a flash. Have you been, doctor?'

'Yes, I have. Where did you stay?'

She paused for a long time.

'I'd like to say Truro but I can't be sure that was the name of the place.'

I knew from Steve's description of the Cornish village that it couldn't be Truro. 'And who were you with?'

'What do you mean?' she asked suspiciously.

'Well, did you go with anyone?'

'No, I just went on my own,' she said coyly.

'Are you sure?'

'Can we speak in confidence, doctor?' she asked, fixing me once again with a deep stare.

'Yes, of course.'

There was a long pause.

'I went with a man I've met recently,' she confessed. 'But I wouldn't want Steve to know. He would be so upset, and to tell you the truth I'm not sure about this new chap either. He can play hot and cold at times. Strangely enough, he's also called Steve. Would you believe it?'

~ o ~

I tried to assess Trish's memory further. She was able to recall events from the distant past well: where she went to school, her first job and the address of the house she lived in with her ex-husband. However, it became clear that she struggled to recall any new information. For example, ten minutes after I gave her a name and address to remember she couldn't recollect any of it. This is a simple clinical test of episodic memory, the ability to recall information from a previous episode in time. It is a

93

cognitive function that requires the hippocampus, a beautifully shaped brain structure that resembles the curved arc of a sea horse, located deep on the inside of the temporal lobe (**Fig. 5**). Hence the Greek origin of its name: hippos meaning horse and kampos a sea monster.

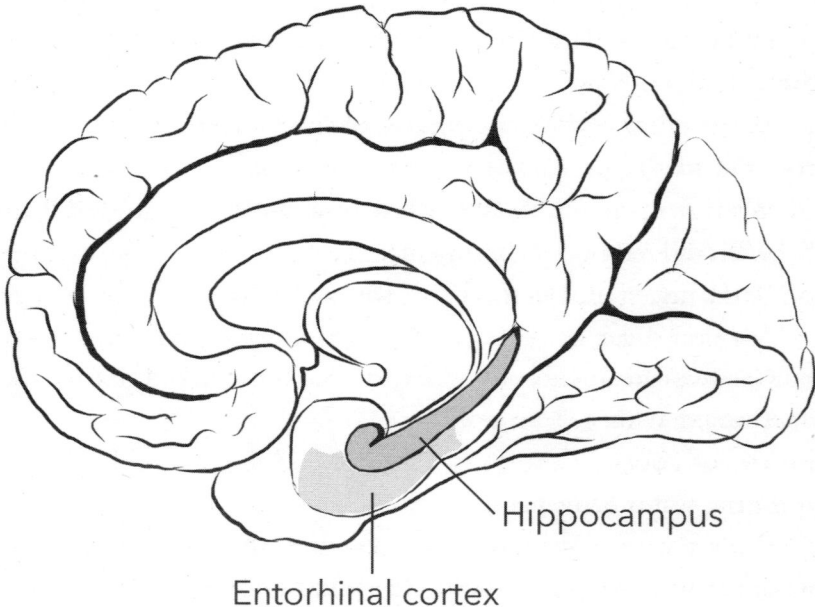

Hippocampus

Entorhinal cortex

Fig. 5 | The hippocampus. A view of the right hippocampus shown from the midline, with the brain structures that lie adjacent to it. These include the entorhinal cortex, which is considered to be the 'gateway' and route through which nerve fibres enter the hippocampus.

Research into the brains of different species has revealed that the hippocampus is an ancient structure which has evolved from the very origins of vertebrates (back-boned animals). In mammals it plays a crucial role in binding together information that belongs to a particular episode in time: the 'what', 'where' and 'who' of a moment that was experienced.[2] What happened? Where did it occur? Who was there? Without the hippocampus, people suffer from severe amnesia. They find it almost impossible to remember new information, but intriguingly they can often recall events from the distant past. They show a marked 'gradient' in their memory such that they find the most recent incidents hardest to recollect whereas older ones are much more firmly embedded.

That damage to the hippocampus leads to deficits in episodic memory was first appreciated spectacularly in the case of patient HM in the 1950s. As a young man, he had a radical surgical procedure performed to treat his epilepsy, which had failed to respond to the drugs available to neurologists at that time. His neurosurgeon at the Montreal Neurological Institute, William Scoville, offered him a new option: resection of the hippocampus on both the left and right side of the brain. The operation turned out to be a success for HM. His seizures reduced in frequency, but it also had unexpected consequences.

After the surgery, he seemed unable to lay down new memories. For example, he would forget that he had been speaking to his doctor minutes beforehand. Years later, Suzanne Corkin, who with Brenda Milner studied HM's memory in great detail over decades, would recount how he would never remember who she was, even though she would reintroduce herself at each meeting.[3] But, just as strikingly, HM (whose name was revealed after his death, to be Henry Molaison) had very little memory of what had happened in the years before his operation. It was almost as if he was living in a permanent present tense.[3]

People like HM raise a question that many of us might never consider: how are memories formed? We remember information so effortlessly, but how? Neuroscientists have been intrigued by how neurons in the brain are capable of storing a seemingly endless number of experiences throughout a lifetime. How do nerve cells create a memory? Everyone is agreed that something has to change in the brain when a new memory is formed, but what is it? How can cells capture the richness of our everyday lives – not only the sights, sounds, smells and other sensory phenomena, but also the mood or the significance of a conversation or event? What exactly is the nature of a memory trace in the brain?

Some researchers have coined the term 'engram' for the neural representation that holds a memory. However, until recently, it has been very difficult to pinpoint engram cells, partly because a memory can be encoded or distributed across a group of neurons.[4] The early scientific studies inevitably had to turn to simple organisms because it was technically easier to record electrical signals from their neurons. Eric Kandel, a Jew who fled to the United States from Nazi-occupied Austria when he was a boy, made some of the pioneering contributions in this area. Of all things, he worked on the humble sea slug.[5] Although the sea slug is much simpler than most mammals, Kandel realised that its nervous system was ideal for studying how new memories are formed – for example, of a potentially threatening stimulus. By performing a painstaking series of experiments over many years, he found that the sea slug's short-term memories were associated with changes in the strength of connections between neurons – at the synapses.

A synapse is the technical term for where neurons meet and communicate with each other (**Fig. 6**). On one side of a synapse is the pre-synaptic neuron. Electrical impulses flow down this nerve cell to reach the synapse – which, effectively, is a gap. For information to cross this gap to the neuron on the other side – the post-synaptic neuron – the electrical impulse is converted to a chemical signal. Neurotransmitter molecules released by the pre-synaptic neuron cross the synapse to reach and bind to receptors on the post-synaptic neuron, triggering an electrical impulse in that nerve cell. This is how nerve cells signal to each other.

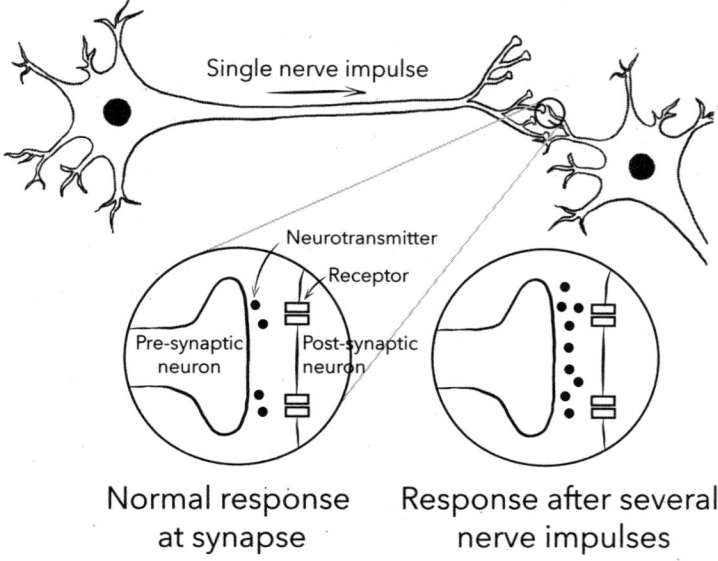

Normal response
at synapse

Response after several
nerve impulses

Fig. 6 | Short- and long-term memories may be due to changes in synaptic strength or increased number of synapses between neurons. Nerve cells communicate with each other via synapses. When a pre-synaptic neuron is activated, it releases neurotransmitter (chemical) molecules. These bind to receptors on the post-synaptic neuron to activate it. Memory is associated with an increase in the strength of the signal across a synapse. Reactivation of the pre-synaptic neuron, even by a single electrical impulse, leads to easier activation of the post-synaptic neuron. This might occur because there is more neurotransmitter molecule release or because of changes in the way the post-synaptic neuron responds.

Kandel discovered that when the sea slug learned a new memory there were changes in the strength of the synapse between specific neurons. In effect, there was an increase in the signal of the information transmitted across a set of particular connections, rather like notching up the gain in an electrical circuit or the volume on a sound system. If it is repeated, the same electrical impulse in the pre-synaptic neuron triggers an impulse in the post-synaptic neuron far more easily.

This finding is precisely what had been proposed by the Canadian psychologist, Donald Hebb, in the 1940s. The changes in strength between synapses was captured by the phrase 'neurons that wire together fire together.' In other words, groups of nerve cells that are active during an event lead to the strengthening of connections between them – and this is the memory trace that's left behind. 'Hebbian' synapses, the term now given to this mechanism, are also considered crucial for 'plasticity' in the brain. With regards to longer-term learning in the sea slug, Kandel found that there were also increases in the number of synapses between neurons involved in creating a memory of a threatening stimulus. This groundbreaking work into memory storage in neurons ultimately led to him being awarded the 2000 Nobel Prize in Physiology or Medicine.[6]

Subsequent research has moved from sea slugs to mice and confirmed the general principles that Kandel discovered. The remarkable Japanese scientist, Susumu Tonegawa, who first obtained a Nobel Prize for his work in immunology in 1990 and then switched fields completely to study neuroscience at MIT, showed how memories could be reactivated in mice. He was able to do this by stimulating specific neurons in the hippocampus that he identified as 'engram cells.'[4] This provided support for the theory that experiences activate a group of neurons which undergo persistent changes, and it is this collection of activated neurons that effectively holds the memory trace or engram. Subsequent reactivation of these neurons induces retrieval of that memory.

Careful assessment of patients like HM have led to the view that, in humans, neurons in the hippocampus play a key role in encoding new information and retrieving relatively recent material. However, more distant episodes in memory might have become consolidated and stored in other brain regions such as the cerebral cortex, so they can be recalled even without an effective hippocampus.[7] This might explain a common observation made by family members of hippocampal patients, who they say have wonderful, sometimes vivid recall of long-past events, but really can't remember what was said to them days, hours or even a few minutes ago. So, the finding that Trish had difficulty with episodic memory, particularly for recent information, pointed to a dysfunction of the hippocampus.

An intriguing feature of Trish's story was that she was not only forgetting information but also mixing up memories from different times. She would confabulate: recount false memories without consciously being aware that those memories were false. Understandably, scientists have become fascinated by why confabulation might occur. What has become clear is that memories are actually not like a photo or a video recording that has captured faithfully what happened. The first professor of experimental psychology at the University of Cambridge, Frederic Bartlett, conducted a large set of investigations into this question and concluded that our memories are effectively reconstructions, not replays, of events.[8]

In one experiment he performed in the early part of the twentieth century Bartlett asked a group of twenty students to read 'War of the Ghosts', a Native American folk story. He then tested their recollection of what they had read at different intervals from minutes to weeks, months or years. Unsurprisingly, his participants forgot details of the story as more time elapsed, but Bartlett also found that they recounted information erroneously, fitting their recollections with their own very Edwardian English cultural backgrounds. For example, some reported that the men in the story were on a sailing expedition at sea to fish, when in fact the Native Americans in the story had been in canoes on a river, hunting seals. In Bartlett's view, their memories were being incorporated into pre-existing 'schemas' acquired on the basis of what they knew. They were effectively recreating memories of what they had read by embedding them into a schema that they were familiar with.

Donald Hebb summarised this neatly. Remembering, he argued, is rather like a palaeontologist reconstructing a dinosaur from an incomplete set of remains found at a dig. The bones and teeth are put together on the basis of prior knowledge of other dinosaur skeletons. Retrieval of memories is an analogous process. Pieces of information are woven into a story by embedding them in past knowledge or a schema of what that event was probably like. So, if a person is briefly asked to sit in an office and then later questioned about what was in that room, it turns out that many people remember seeing objects that fit their schema of what an office contains, such as books or pens, even if it was the case that there were none of these items in the particular 'office' they had been in.[9]

Psychologists have also found that memories can be fragile for a different reason: our recollections are vulnerable to suggestion. They can be reconstructed differently depending on the way a question is posed. In a famous experiment, Elizabeth Loftus and John Palmer got students to watch films of two cars colliding and then immediately afterwards asked them questions about what they had seen.[10] The critical question concerned participants' estimates of the speed of the vehicles. Some of the volunteers were asked, 'About how fast were the cars going when they smashed into each other?' But for others, the verb 'smashed' was replaced by 'collided' or 'bumped' or 'contacted.'

Recollections of the speed of the cars varied with the verb used in the question, with the highest estimate (40.5 miles per hour) when 'smashed' was used and the lowest (31.8 miles per hour) when the question included the verb 'contacted'. A week later, the students were asked whether they had seen any broken glass in the movie. Those who had originally been asked about the speed of the cars with the verb 'smashed' were far more likely to say 'yes' in response to the question, even though broken glass was not present in the film. Their reconstruction of the event a week later had been further distorted by the nature of the question posed immediately after it.

Such distortions can have devastating consequences in real life.

False memories can be created by the way a question is phrased. Eyewitness testimony has often been found to be unreliable in court cases and such unreliability can be emulated in psychological experiments. In one study, people watched an eight-second security video and were subsequently asked to pick the gunman from photographs. Every observer selected someone, despite the fact that none of the people in the photos was the actual gunman.[11]

All of these examples demonstrate the fragility of normal human memory and reveal how we all can misremember. One way to think of confabulation in patients is that it is an extreme form of how memories can be distorted in healthy people. In cases of spontaneous confabulation, such as in Trish, the patient is convinced of the truth of their stories. Indeed, the content can often be traced back to real events in their life or to what they have observed – for example, on the television. Some investigations of patients who confabulate suggest that they have considerable difficulty in distinguishing between memory relevant to ongoing reality and memory of previous, irrelevant experiences.[12] More generally they find it hard to distinguish the sources of their memories; they find it difficult to know when in time each memory comes from – a 'source monitoring' deficit. Trish, for example, had muddled up events from the distant past (President Nixon and the Watergate scandal) with what was happening now (in Syria). Other research suggests that when a patient confabulates, the normal processes that are involved in assessing the quality of a retrieved memory are dysfunctional. Instead of finding it implausible that a series of events actually happened, the patient considers it sufficiently credible to articulate it. Either mechanism might have accounted for Trish's confabulations.

~ o ~

Were any brain regions apart from the hippocampus affected, or for that matter working normally, in Trish? Unlike Michael, discussed in the previous chapter, her semantic memory appeared largely intact. She knew the meanings of words and could name objects without difficulty.

Psychologists also recognise a third type of memory system – short-term memory. Quite unlike episodic memory, this decays over just a few seconds unless the information is rehearsed (repeated) rather like when we are trying to retain a telephone number that someone has given us but don't have anything to write it down or record it with. We performed some further tests and found that Trish's short-term memory for numbers (a test of verbal short-term memory), as well as for spatial locations (visuospatial short-term memory) was impaired. For example, she could barely remember more than a three-digit sequence. Similarly, if I tapped out with my pen a sequence of locations on my desk, she could reproduce this correctly – but only if the sequence involved two or three locations, no more.

These different types of short-term memory impairment are often indicators of parietal lobe dysfunction.[13] Verbal short-term memory impairments occur as a consequence of *left* parietal damage, whereas visuospatial short-term memory errors are associated with defective *right* parietal function (**Fig. 7**).

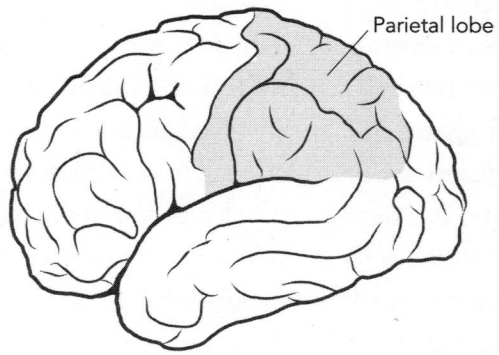

Fig. 7 | The parietal lobe. The parietal lobes in each cerebral hemisphere are specialised for different functions in human beings. Left parietal functions include verbal short-term memory, calculation and praxis, whereas the right parietal lobe is more specialised for visuospatial and attentional functions.

But it turned out that it was more than just her short and episodic memory that Trish was having difficulty with. She struggled to do any calculations apart from simple addition, whereas previously she had apparently been very numerate. Such dyscalculia, like verbal short-term memory deficit, is also often an indicator of left parietal dysfunction. Next, I tested Trish's praxis by asking her to copy some meaningless gestures I made with my hands. She was unable to reproduce the configuration of the fingers I showed her with either hand, demonstrating that she had limb apraxia (**Fig. 8**). This, too, is a test of left parietal lobe function.

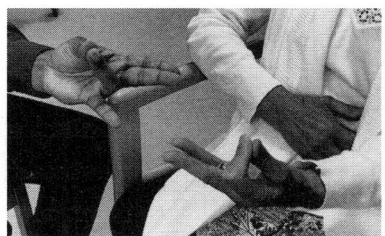 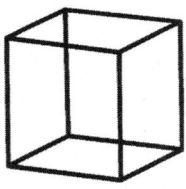 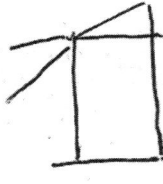

Fig. 8 | Limb apraxia and constructional apraxia. The left panel shows impaired copying of meaningless gestures – limb apraxia. The patient is attempting to copy the configuration of fingers demonstrated by the doctor (on the left). The right panel shows her copy of the drawing of a cube, revealing constructional apraxia. From Tabi & Husain (2023).[13]

Finally, when I asked Trish to copy some simple line drawings, she found the task very confusing, making several attempts to reproduce the figures, but with none ending in success – a phenomenon that is termed constructional apraxia (**Fig. 8**), which is associated with right parietal cortex disorders. In contrast, when I gave Trish tests which assess functions of the frontal lobe she performed relatively well.

Studies of the casts of skulls of our ancestors have suggested that the parietal lobes expanded enormously in evolution. Some researchers have proposed that this is linked to two important skills that modern humans have acquired.[14] The first is praxis and the ability to configure our multi-jointed hands to both make and use tools. The implements that humans developed have become ever smaller over time, sometimes requiring a completely new set of actions to manipulate them properly. From flakes of flint used to cut meat in the Stone Age to modern watchmakers' screwdrivers and the precision tools used for microscope-assisted surgery requiring fine control of our fingers, we have come a long way in how we fabricate and use tools. Parts of the *left* parietal cortex, it has been argued, became highly specialised for the manual control required to both skilfully craft and employ the implements we invented, a cultural development that was crucial to the success of our species.

The second important skill that humans have developed is that of being able to make plans for new constructions, to draw designs not only of dwellings and buildings, but also clothes, means of transport (from carts to rockets) and a host of objects, including tools. We are able not only to sketch plans or draw precise blueprints for ourselves, but we can also convey these in a visual format to other humans, on paper, digitally or using scaled models. Being able to make such plans in two or three-dimensions requires 'visuoconstructive' skills, which we now know are an important function of the *right* parietal cortex.

Short-term memory, both verbal and visuospatial, together with the ability to perform calculations, are also considered to be central to the cultural transformations that have occurred in human evolution. All these cognitive processes – praxis, visuo-constructive ability, calculation, verbal and visuospatial short-term memory, to which the parietal cortex makes such a fundamental contribution – have allowed us to solve complex problems, both real-world and abstract. We have the power to imagine how to develop new and ever more complex ways of constructing not only tools, but machines and large-scale structures, such as skyscrapers, cruise ships and spacecraft. And we can communicate to others how we might do this by creating detailed plans. Small

104

wonder that it has been argued that many important aspects of human material culture have depended on the development of the parietal cortex, with different forms of specialisation of function occurring on the left and right sides of the brain.[14]

The examination of Trish's cognitive functions in the clinic pointed to dysfunction of the hippocampus (episodic memory impairment), as well as that of both left and right parietal cortex (short-term memory deficits, dyscalculia, limb apraxia and constructional apraxia). This is how neurologists go about attempting to 'localise' which parts of the brain might not be functioning properly and *where* in the brain the damage might be occurring. Having done this, the next step was to turn towards thinking *why* this was happening. What sort of pathological process might be underlying this pattern of brain dysfunction in Trish?

To a neurologist, the combination of hippocampal and bilateral parietal dysfunction, in the context of a slowly progressive history of four years, suggests a neurodegenerative condition that has a predilection for these specific brain regions. Top of the list of such disorders in an elderly person is Alzheimer's disease, where the pathology often appears to start just outside the hippocampus, in the entorhinal cortex. Lying adjacent to the hippocampus and often considered to be the 'gateway' to it because nerve fibres from all over the cortex converge through it (**Fig. 5**), the entorhinal cortex seems particularly vulnerable to the brain changes that occur in Alzheimer's. Soon this pathology spreads into the hippocampus and, if there is sufficient damage, it leads to episodic memory impairment. Then the pathology spreads further to affect the left and right parietal cortex, in turn leading to short-term memory deficits, dyscalculia, limb apraxia and constructional apraxia.

But Trish was in her mid-fifties. Could she have developed Alzheimer's so early? If so, why? These were exactly the questions I asked myself as I completed the rest of the neurological examination. I looked for clues to suggest an alternative diagnosis, but there were none.

~ o ~

Auguste Deter was about the same age as Trish when she also came to medical attention. It was 1901. Her husband, Carl, a railway clerk, had noticed for several years that her memory was not as sharp as it used to be, but her symptoms had suddenly taken a turn for the worse. Auguste started to become very suspicious. She was jealous. She became paranoid that her husband was having an affair, a fear which proved to be unfounded. Her delusions then extended to her neighbours who, she thought, were trying to kill her. Her memory also seemed to be deteriorating rapidly. In desperation, her husband took her to the Institution for the Mentally Ill and for Epileptics in Frankfurt, where she was admitted by one Alois Alzheimer, a 37-year-old psychiatrist who had a particular interest in neuropathology (**Fig. 9**).

Fig. 9 | Alois Alzheimer and his famous patient, Auguste Deter.

Alzheimer's clinical notes on Frau Deter were lost for many decades but were amazingly rediscovered by chance in the 1990s. They reveal just how poor her recollection of details of her personal life were. Alzheimer became fascinated with Auguste because she had developed dementia at such a relatively young age. Indeed, when in 1903 he decided to move from Frankfurt to Munich to take up a job offer under Emil Kraepelin, perhaps the foremost psychiatrist of the time – and often considered as the founder of modern scientific psychiatry – Alzheimer asked that he be informed when Frau Deter died, so that her records and brain might be sent to him after the post-mortem.

In 1906, after Auguste had passed away, Alzheimer reported his neuropathological findings at a meeting of psychiatrists in Tübingen, Germany. His talk received little attention, but in it he described two very unusual features of what he had seen under the micro-scope (**Fig. 10**). First, outside of the nerve cells – the neurons – he observed deposits of a peculiar substance, which we now call 'plaques' and know to be made up of the protein amyloid. Second, within many of the neurons there were tangled bundles of what seemed like tiny fibres. Nowadays we refer to these as 'tangles'. They are made up of a different protein called tau.[15]

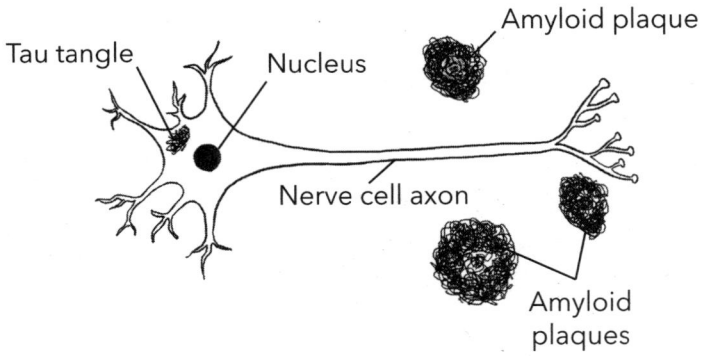

Fig. 10 | Plaques and tangles in the brains of people with Alzheimer's disease. Under the microscope, amyloid plaques outside neurons and tau tangles within them can be seen in the brains of people with Alzheimer's disease.

Alzheimer's findings might never have received any recog-nition had it not been for his boss, Kraepelin, who in his 1910 *Handbook of Psychiatry* generously gave the condition that

Auguste Deter had suffered from the name of 'Alzheimer's disease'. Alzheimer himself sadly fell ill while travelling in 1912 to take up the post of Professor of Psychiatry at the University of Breslau. He never really recovered and died a few years later in 1915. His legacy, however, remains strong because the microscopic features he described have been confirmed as the hallmark of the disease named after him. Moreover, cases of young-onset Alzheimer's disease, like Auguste Deter, sparked great scientific interest when it became clear in the early 1990s that there were rare cases of Alzheimer's disease which run in families. In these unfortunate people, the disease manifests in their forties and fifties, and is associated with mutations in genes that strikingly lead to deposits of amyloid plaques in the brain.

With this knowledge came the development of the hypothesis that the laying down of amyloid triggers a cascade of molecular events in the brain, ultimately causing the death of neurons and, with the passing of time, dementia.[16] This new proposal in turn led to renewed interest in research on Alzheimer's disease, including the development of techniques to measure amyloid and its co-culprit tau. First came the ability to detect these two molecules in the cerebrospinal fluid, the liquid that bathes the brain. This can be 'tapped' by performing a lumbar puncture: a procedure which involves inserting a needle into the lower back to draw off some of the same cerebrospinal fluid from around the lower spine. More recently, it has also been possible to measure amyloid and tau in the blood of patients suspected of having the condition. In some specialised centres, it is now possible to use new types of scan which involve injecting a radioactive substance that binds to amyloid or to tau in order to assess more directly in the living brain whether these molecules are present in excessive amounts.

These methods have begun to revolutionise how the diagnosis

of Alzheimer's disease is made, not only in younger-onset cases but also in the more common cases observed in older people. The proposal in older people is that, although a specific gene might not be to blame in these individuals, they too have amyloid deposits in their brains and that this phenomenon is the trigger that leads to the loss of neurons and dementia. Modern-day scanning techniques have also allowed us to track the progression of the pathology in Alzheimer's disease, which typically moves from the entorhinal cortex, through the hippocampus and then on to the parietal cortex.

~ o ~

I explained to Trish that we needed to perform some tests, including an MRI scan – which would show us the structure of her brain – and a lumbar puncture to examine the cerebrospinal fluid. She wasn't so keen on the latter but agreed to have it done. The MRI scans were available three weeks later. When I looked at them on my office computer they were clearly abnormal. Just as the cognitive assessment had suggested, Trish's hippocampi and parietal lobes on the left and right side were smaller compared to the rest of her brain. They had seemingly shrunk, which is very characteristic of the pattern observed in Alzheimer's disease. Her cerebrospinal fluid analysis confirmed my suspicions too, revealing abnormal levels of amyloid and tau, the signature findings of what Alois Alzheimer had observed in the brain of his famous patient, Auguste Deter.

In view of these results, I went on to request a fluorodeoxy-glucose PET (positron emission tomography) brain scan; this reveals if parts of the brain are not using glucose – the fuel that neurons require to function. Trish's scan showed that, beyond the areas which were smaller on the MRI scan, there were regions in the temporal and parietal cortex which were not taking up glucose normally, indicative of brain regions that were not functioning appropriately even though they appeared normal on the structural MRI scan. All these investigations supported my clinical impression and pointed consistently to a diagnosis of Alzheimer's disease.

Breaking the bad news proved to be surprising. Trish remained defiantly upbeat even when I showed her brain scans.

'Well, I'm not going to let that kind of thing get me down, doctor. There's a lot more of life to live and we don't know for sure I've got Alzheimer's, do we now?' she concluded.

'It's true that we can't tell with one hundred per cent certainty,' I responded. 'To do that we'd have to look at a little bit of your brain under the microscope, and we're not going to do a brain biopsy because that could do more harm than good. Nevertheless, all the assessments and scans you've had do make Alzheimer's the most likely diagnosis, I'm afraid,' I explained.

Steve was both upset and relieved.

'I've been frightened that this was the cause for a long time . . . but at least we have an explanation now.' There were tears running down his face. 'What about treatment? Is there anything? I've been reading about this drug that clears amyloid protein from the brain. Could Trish try some of that?'

Clearly, Steve had been researching the possibility of Alzheimer's even before I had given them the diagnosis. He'd learned about new experimental treatments that were using antibodies to bind to amyloid and promote its clearance from the brain. Some of the earlier trials had failed to show that there were any benefits to such 'mopping up' of amyloid on the cognitive functions of patients with Alzheimer's disease. Although the drugs did clear amyloid, there had been no real impact on a patient's memory or other cognitive abilities. Nearly a decade later, more promising results were to emerge from clinical trials using this approach but with different drugs, although the benefits reported were controversial. But at the time when I was seeing Trish, there was no evidence that any drug made any material difference to the trajectory of Alzheimer's disease, nor were any licensed to be used in such a way.

'I'm afraid there isn't an amyloid clearing drug that is approved for use yet. But we can start you on a drug that can make a small difference to your memory, Trish.'

I was referring to a medication called donepezil that didn't alter the progression of the disease but did boost the levels of acetylcholine, one of the brain's neurotransmitters (the chemicals used by nerves to signal to each other). Clinical trials had shown that it could boost cognitive performance, but only modestly in patients with Alzheimer's disease.

'We should also discuss genetic tests,' I continued, 'to see if we can discover why you might have developed Alzheimer's at a young age.'

I went over what a positive genetic test result might entail. It could provide an explanation for Trish's condition, but it also had implications for her family, including her children, because it could mean that they too were going to develop the disease. We had learned of these implications from a remarkable extended family in the state of Antioquia, Columbia. This kindred comprises approximately five thousand people, of whom some one thousand five hundred are known to carry a mutation in the *PSEN1* (*Presenilin-1*) gene that is inherited in an autosomal dominant manner. This means that anyone carrying a single copy of the mutated gene will develop Alzheimer's with one hundred per cent certainty.

The protein that the gene codes for affects how amyloid is handled by cells; mutations in the gene lead to increased levels of amyloid aggregating to form plaques in the brain, which ultimately result in dementia. Tonegawa's research team at MIT had demonstrated that it was possible to improve memory recall in mice who had mutations in the very same gene by activating

111

the hippocampal engram (memory) neurons.[17] In addition, their experiments revealed, quite astonishingly, that it was possible to plant false memories by stimulating these neurons,[18] a scenario which is perhaps analogous to the phenomenon of confabulation in patients. With the advent of genetic testing, it was possible to discover in the Columbian family precisely who was going to develop Alzheimer's, and who wasn't, years before they would normally develop symptoms. The same, I thought now, might also apply to Trish's children.

'But my parents didn't have Alzheimer's,' she told me. 'How could that be if I have a genetic condition?'

'Well, we don't know yet whether you have a genetic mutation,' I said, sidestepping the question.

Trish's father, it turned out, had died in his forties in a car accident, so it was possible he might have gone on to develop the same disease as his daughter now had. In the Columbian family, some research suggests that the mutation might have been brought to South America by a Spanish conquistador in the 1600s, but the exact nature of the genetic causes of young-onset Alzheimer's disease in that country has turned out to be more complex than originally thought. Another mutation in the *PSEN1* gene seems to have been brought to Columbia from Africa, possibly by someone who was enslaved.

'I don't think I want my kids to know if they're going to get this . . . if it really is Alzheimer's,' Trish responded, maintaining her scepticism. 'So, I'd rather not have any genetic tests, thank you.'

We didn't pursue that line of investigation further. I could see that Trish was in some degree of denial about her diagnosis. There was no point in pressing it home. I simply gave her a pamphlet which provided some information on Alzheimer's disease and where she might find reliable information, including from counsellors who work with young dementia patients.

'What about your work?' I asked.

'What do you mean?' Trish responded, the note of defiance clear in her voice.

'I think it might be important to have a discussion with your employers, especially if you've been making mistakes. Sooner or later one of those could be serious, so—'

'The mistakes have been a few minor things. I'll think about it,' Trish interrupted, making light of matters and clearly not wanting me to pursue this.

'I'm afraid I'm also going to have to ask you to inform the driving agency and your motor insurance company about the diagnosis. They might write to me for more details.'

'That as well!'

'I'm sorry, but this is a legal requirement.'

I gave Trish a prescription for donepezil and arranged to see her in a few months. Before I had finished explaining when we would reassess her, she asked: 'By the way, do you have the results of my brain scan yet?'

This is exactly what Steve had meant about repeating questions and forgetting new information over a period of a few minutes. I went through the test results again. As she left, Trish shook my hand, but she clearly wasn't convinced there was a problem or by the diagnosis I was making. I didn't feel confident that she understood the full significance of what I had said.

'If you – if either of you – have any more questions, please do send me an email,' I said. 'I'd be very happy to answer them. We could arrange a call.'

'Thank you, doctor,' said Steve. And then, 'Let's go and have a chat over coffee,' he suggested to Trish, trying to be comforting.

'It'll be a fine hot chocolate for me, Rockefeller,' she insisted with a wicked smile. 'I'll not be short-changed on that.'

I was obviously looking bewildered.

'Trish has a weakness for cocoa from a specialty shop in Soho,' Steve explained.

'It's not a bloody weakness. It's a strength,' Trish retorted. 'He really doesn't understand hot chocolate, doctor. It's medicinal.' She chuckled. 'It'll be my cure, that's for sure.'

~ o ~

Leaving the hospital that evening, I wondered how the conversation had gone between Trish and Steve when they got to the cocoa shop in Soho. More importantly, how would they cope over the next few months, she seemingly in denial and in such good spirits, he obviously exasperated but trying to be supportive. The bus I was on was going past the Physic Garden on its way towards Sloane Square in Chelsea. From the top deck it was difficult to make out any details. Darkness had cloaked the gardens in a deep indigo inscribed with vague, darker dendritic forms in the middle distance. First established by the Society of Apothecaries to grow medicinal plants, the Physic Garden is a haven in modern London, but in the past, it was an extremely busy place through which plants from around the world were introduced to Britain.

I smiled. Thinking about Trish and her hot chocolate while passing this place led to the peculiar recollection of Sir Hans Sloane, one of eighteenth-century London's most famous physicians, and the great benefactor of the Chelsea Physic Garden. It was he, too, who has often been credited with bringing chocolate from Jamaica in a form that was appreciated by the English palate. By mixing it with hot milk and sugar, Sloane supposedly introduced hot chocolate to London, a beverage which – as Trish herself surmised – was believed to have healing properties. Certainly, apothecaries at the time set great store by its medicinal attributes, leading to the spectacular rise of the drinking of hot chocolate in Britain.

Sloane himself was busier with ambitions far grander than hot chocolate. Born into the very modest home of a tax collector in what is now Northern Ireland, he most remarkably went on to become physician to three English monarchs – Queen Anne, George I and George II – as well as being president of both the Royal College of Physicians and the Royal Society. His wealth grew thanks to his marriage to Elizabeth Langley Rose, who had inherited profitable sugar plantations in Jamaica, and with this money Sloane had set about collecting all sorts of plants and objects from around the world. Ultimately, his legacy, undoubtedly built on money made from slavery, led to the birth of no less than the British Museum, as well as to a host of London streets and locations – including Sloane Square – being named after him.

Despite how he made his money, Sloane did have an astonishing career, breaking into the inner circles of London society and successfully becoming a leading figure within those circles when he was from a very modest background in Ireland. Indeed, some historians conclude that since his provincial Irish brogue would have been considered quite barbaric in polite society, he must have relinquished his accent very rapidly in order to make the extraordinary ascent he did. He was an outsider who knew how to operate, so well in fact that he became a leading insider. As Isaac Newton wryly commented, Sloane was 'a very tricking fellow'.[19]

My bus had now reached the Embankment. The glistening black expanse of the Thames was illuminated by the graceful arcs of lights suspended over the majestic Albert Bridge. It was time for me to get off. I wondered how often Trish was going to be sipping high-end hot chocolate in Soho from now.

~ o ~

I next saw Trish four months later. When I spoke to Steve on his own, he thought that perhaps there had been some improvement after we had started her on the donepezil medication, but it was mild. Trish was still experiencing difficulties remembering important information, and this was becoming even more evident at work, he said, where her employers were very concerned about her continuing. There was pressure on her to resign. Life outside work was also changing.

Trish had a lively, close group of friends but she was now seeing less of them. It was apparently her decision rather than theirs. Steve thought it was because she was finding it increasingly challenging to follow their conversations. He'd noticed when the friends had visited their home that Trish would often have no recollection of the people or events the group was discussing. She appeared to be lost, bewildered in the midst of the fast-flowing chat and clearly feeling excluded, whereas previously she had usually been the focus of attention, the person cracking the jokes. It seemed like she might have been confabulating with them too. Sometimes she would say things that they looked puzzled by. But Trish wasn't one for pity and had decided not to mention anything about the possible diagnosis. However, Steve had noticed that she was now feigning other commitments when her friends organised nights out and was slowly losing her connections with them.

Worse still, Trish had succumbed to a scam telephone call from someone supposedly selling timeshare properties in Spain. She had lost over £8,000 by handing over her bank details. Steve was fighting to get this money back, but her children were very annoyed. They'd accused Trish of being incompetent and had argued fiercely with her. Now they weren't speaking to her, but Trish still hadn't told them anything about her diagnosis of Alzheimer's disease. She wanted to remain independent, even if this meant becoming estranged from her family. Steve was finding it more and more difficult to cope with the situation.

'To be frank, doctor, I'm not sure I can continue in this way. She just won't listen to me when I say we have to tell people so that they'll understand. I can't really carry on with her if she doesn't.'

Steve was clearly at the end of his patience. If Trish were to lose him, she would become even more isolated. When I brought her into the consulting room to speak to her alone, it was evident that she was still persisting with shrugging off the diagnosis.

'To be honest with you, I don't understand what all the fuss is about,' she said. 'People forget, don't they?'

'They do.'

'Then we're agreed,' she interjected. 'I'm wasting your time, aren't I now?'

'No, that's not what I think at all,' I responded quickly. 'You're definitely not wasting my time. Everybody does forget, but I'm afraid you're forgetting a lot more than most people, and it's not just your memory that seems to be affected.'

It is difficult to be blunt, but it's necessary if someone doesn't seem to be acknowledging they have a problem. Trish's face narrowed.

'How do you mean?' she asked accusingly. She was clearly worried that Steve had said things that she'd rather he hadn't. I wasn't going to avoid the issue.

'Look,' I said, as calmly as I could, 'I've heard about the fact that you've also made some bad decisions lately, like the money you lost in the scam.'

'That can happen to anyone!' Tris retorted.

'It can,' I said softly. 'But it hasn't happened to you before.'

'Well, there's always a first time for everything, isn't there?' she said, still defiant.

'I think the reason for many of the things that have been happening to you – at work, with your group of friends and with other people – might have something to do with the fact that you have Alzheimer's disease,' I said slowly.

The room fell into silence. Trish turned away. For a moment I thought she was about to storm out. But she looked back at

117

me, her eyes glassy. There was a little opening of her lips. Then she began to sob, quietly.

'I know. I've just been so frightened,' she explained, tears now coursing over her hollow, pale cheeks.

I went to pass her a box of tissues, but she took my hand and held it tight. We didn't say anything for several minutes.

'I've known there must be something wrong for a very long time, doctor,' she said at last. 'But I've been . . . so worried that Steve and everyone else would abandon me if they thought I had dementia.'

I was startled by her explanation. I hadn't appreciated why she had been so strong-willed about not accepting a diagnosis. Now I knew. She was scared and I hadn't realised. I felt awful, inadequate, and moved by how Trish had hidden her true sentiments.

'They're not going to leave you if they understand what the problem is.'

'No?'

'No. It's not your fault you have this condition. But you have to be honest with the people who love you. They care for you but they have to understand what you have. And you might have to listen to what they say, especially Steve. He's trying his best but he's finding it very difficult because you don't seem to acknowledge there is a problem.'

'Thank you for telling me,' she said, nodding her head. 'I know I will have seemed really ungrateful, but I do appreciate all that you've done. Thank you for telling me the truth.' She gave me a tight squeeze of the hand, then wiped away the tears from her face.

Trish's response in coming to terms with her diagnosis had been denial, but this very act was threatening to make her situation worse with the people who loved her the most.

When Steve came into the room, he was alarmed to see how

distressed Trish looked, but she smiled reassuringly, got up and gave him a very long hug.

'It's alright, I know I have Alzheimer's and I also know that you're amazing, doing all the things you do for me,' she announced to Steve, who couldn't disguise how taken aback he was.

'Let's go for some cocoa,' she said with a smile.

It turned out that admitting there was a problem helped everyone, including Trish. She stepped down from her job, which had become hugely stressful for her. When she explained her diagnosis to her children, they were apparently both shocked and ashamed about the way that they had behaved. They became far more sympathetic and supportive of their mother. Trish's friends, too, were also very understanding once they realised why she was so different. They gave her space to talk in their conversations and found that if she was given sufficient time, her sense of humour could still be extremely funny. I had also learned a lesson. Denial of a diagnosis, especially one that is going to change someone's life, isn't simply about being obstructive. It can be due to a very real fear of what that diagnosis will mean for the relationships we hold most dear.

4

Visitors in the night

'It must have been very frightening for you.'

'What must have been, doctor?' asked Wahid in his clipped Punjabi accent.

'Well, your GP says in her letter that you thought you saw some people in your bedroom in the middle of the night.'

He hesitated, glancing away. 'Doctor, I think I must have been confused when I was speaking to her,' he blurted out.

'Confused?' I enquired, studying him intently. Wahid Razzaq was a tall, lean man, still looking very athletic in his late fifties. He sported a well-trimmed moustache and beard, now greying in places, but meticulously attended to with obvious care and precision. He wore a uniform of some kind, a blue jacket and tie with a red insignia, but I wasn't sure what it was for.

'Yes. I don't know why I said that to her. I must have been dreaming.' The words tumbled out without conviction.

'I see, so why are you here, Mr Razzaq?' I asked pointedly.

'I got the appointment, sir,' he explained, opening his hands outstretched in front of him as if to ask what else could he do in response to the summons.

I decided to change the subject.

'You really don't need to call me sir. Where were you born, by the way?'

'Born? Oh, I am from Lahore. Have you ever been?'

'No, I'm afraid not. But I hear from my parents, who spent some time there when they were young, that it is a beautiful city.' In fact, I thought, it was actually the city where I had been conceived.

'Indeed. It is the city of Mughal emperors. We are very proud of it.' He smiled, now seeming more at ease. 'It is a wonderful place. Sometimes, I am so sorry to have left it.'

Lahore had indeed been the centre of the Mughal Empire, a thriving city for literature, art and architecture in the sixteenth and seventeenth centuries. Years later, it was to become a key city for the British administration of its empire in India.

'And where are you from, if you don't mind me asking, doctor?' he continued.

'I was born in East Pakistan, when it was called that. Now, of course, it is Bangladesh.'

'Never mind,' said Wahid. 'We still accept you! Once a Pakistani, always a Pakistani,' he added with a grin.

Although this was presumably a heartfelt joke, Wahid would have very clearly understood that, in reality, once a Pakistani didn't mean you always remained a Pakistani at all. The state of Pakistan had been created in the rushed, some might say bungled, dissolution of British colonial rule of India in 1947. What was then British India had been partitioned so that Muslim communities in the east and west of the region would live in a single state. But the two parts of this state – East and West Pakistan – were separated by nearly one thousand miles, with the modern country of India (which has a Hindu majority) created between them. The entire process of partition was an artifice with tragic consequences for those caught up in the religious violence that it triggered. In the ensuing chaos, up to ten million people

121

migrated across the subcontinent, with many Muslims fleeing to the newly formed areas that were under Pakistani rule, and millions of Hindus making their way to India.

My father had been born in Bengal at the time of the British Empire and grew up in the region that became East Pakistan. But to receive his training as a radiologist he had flown nearly a thousand miles with my mother to live in Lahore, situated in West Pakistan. That is how they had found themselves in the city where Wahid had grown up. East and West Pakistanis, although all mostly Muslims, were from completely different cultural backgrounds. Bengalis living in the East spoke an entirely different language. They had separate social, culinary, musical and literary traditions from people in the West. Inevitably, following a violent war of independence in 1971, East Pakistan split away from West Pakistan to become present-day Bangladesh.

One of the legacies of the British Empire was the emigration of people from all over its former colonies. Over a million people of Pakistani origin alone now live in the United Kingdom. Like many Punjabis – just one of the ethnic groups from Pakistan – Wahid had come to London in search of work in the 1960s. Life had been extremely difficult for him.

'Nobody wanted us. We'd get told to take our stinking curries and go home. Eventually, I found a job sweeping the streets, but that wasn't safe. People would spit on me or attack me – apparently for taking their jobs. I couldn't say "But none of you want to do this job! That's why I have it" because that would get me into even more trouble.' Eventually, over thirty years ago, he had managed to become a bus driver, and he had stuck to it.

'You enjoy your job?'

He sighed. 'It is very important to me.'

'That's good to hear.'

'Yes, it is, how do you say, my livelihood. Without it, I would be finished.' He rubbed his palms over each other in a single flourish to make the point.

Suddenly, Wahid's head tilted back, and with his deep brown eyes wide open in seeming disbelief, he stared at the corner of the ceiling, clenching his teeth.

'Are you alright?' I asked as I walked around to the chair he was perched on.

'I'm really fine, doctor.' He turned his gaze towards me, his face quite gaunt.

But Wahid wasn't fine. He was still glancing periodically up at the corner.

'You look worried. What can you see up there?'

Wahid shook his head. 'I can't see anything.' He paused. 'Can you see anything there, doctor?'

'Look, I am here to help. I can't see anything odd in the corner, but you looked frightened.'

Wahid's hands rose to his high cheeks and moved slowly down, stroking his greying beard. He shook his head again. There were now a few beads of sweat on his forehead.

'No, I can't really tell you, sir.'

'All that you say here is confidential between us, Mr Razzaq,' I explained, trying to reassure him.

Once more, he stroked his beard with his right hand.

'Really?' Wahid's top lip moved forwards. He hesitated.

'Yes, and I might be able to help, but I can't do that unless I understand what the problem is.'

Wahid considered what I had said for a long time. He sighed again and then came his confession.

'I have been seeing people, sometimes animals. They come and go. Sometimes they last for seconds and are gone. Other times they can be there for many minutes. It is worst in the evening or at night. I have woken up to see a group of hooded men looking down at me as I lay in bed. They just stared. It was very frightening the first time. I don't understand what is happening.'

'What do you see exactly?'

'They are like shadows, dark shapes, but if I look closely, they are people.'

'And do they ever speak to you?'

'What? Oh no, thank God. No, they don't say anything.'

'Do they ever put thoughts into your mind?'

'Oh no. They don't. They just stare.' He paused. 'Sometimes I have seen things that look like mice or large spiders running over the floor.'

'How often do you see them?'

'Perhaps once a day. I thought I saw a face in that corner up there,' he said, pointing to the ceiling.

'And did you?'

'Yes, I think so, but he was gone very quickly.'

'Do you ever recognise any of the people?'

'Not really. They seem very familiar now because the same ones come back, but I don't know them.'

'And you don't think they are real?'

'Oh no, of course they're not real, doctor.'

'Have you ever seen them when you are driving the bus?'

Wahid's upper lip moved forwards again. 'Yes, once, and that is what I am worried about. I saw the dark shadow of a man running to my left, but I knew he wasn't real. It was only for a few seconds, but it makes me worried.'

'What are you most concerned about?'

He held his head in his hands. 'That I am going mad.'

I could understand Wahid's reluctance to tell me about the unusual visitors that came to him in the night. Madness, as the medical historian Roy Porter observed, is probably a condition that is as old as mankind.[1] Although defined in many different ways throughout the centuries, one consistent theme has been the 'othering' of people considered to be mad by the rest of society. They are different, often visibly so, and because of this they can be ostracised by people around them. But what Wahid had described up to now were only visual hallucinations – in other words, perceptions experienced in the absence of any external stimulus. Visual hallucinations, on their own, are not sufficient to consider someone to be mad, at least not in Western medical culture.

Wahid had not mentioned anything else to suggest either psychotic or disorganised thinking. He showed no evidence of harbouring any delusional thoughts – false beliefs that are firmly held on inadequate grounds and that would be unexpected on the basis of someone's cultural background or education.[2] He also had insight into his symptoms and their potential significance if people were to find out that he was experiencing hallucinations. None of this pointed to a serious psychiatric diagnosis. I let Wahid speak on about his concerns. As I listened to him talk, it became clear how madness means different things to people from different backgrounds.

'In my community, madness is not understood. It is not acceptable, doctor. People think that you must have done terrible things for this to happen to you,' he explained. 'Some of them think you are possessed by a *jinn*.'

In some cultures around the world, hallucinations represent a form of demonic possession of the person and the mind. Among Pakistani people, and many other Muslims around the world, there is a belief in jinn (more familiarly referred to as genies in Western literature). Considered to be invisible beings (the Arabic meaning of the word jinn is 'to hide'), who live alongside humans as snake-like or other creatures, jinn are able to change their form at will, and crucially can take possession of human beings. By entering a person's body, they can control an individual or alter the way they behave, even change their identity.[3] Some

Muslims take such possession to be a manifestation of a punishment from God. In South Asian communities in the United Kingdom, many people not only believe in jinn (80 per cent according to one survey), but also consider them to cause mental illness (58 per cent).[4] To be cured requires the expulsion of the jinn from the body.

In part because mental illness carries with it this burden of being possessed by an evil spirit, it is also associated with a huge social stigma in these communities. Little wonder that many in and from Pakistan overtly deny experiencing mental health symptoms and do not seek medical help. Instead, their recourse is to pray or to seek out faith healers willing to perform exorcism. Perhaps these beliefs contribute to the fact that in 2020 there were only five hundred psychiatrists in Pakistan caring for patients in a population of 200 million.[5] Even in the UK, people originally from South Asia access mental health services with far less frequency than the indigenous population.[6]

It is perhaps paradoxical how the concept of being possessed by an external force, such as a jinn, is an accepted explanation of mental illness among some communities. For Western psychiatrists it would be considered to be a delusional belief if a person from a non-Muslim background were to voice such a view. So, a patient's cultural background matters when making a diagnosis. Symptoms such as visual hallucinations can have negative religious or spiritual attributions in some cultures. This may mean that those communities are less likely to be sympathetic to a sufferer than they would be if the illness was considered to be a physical one affecting the brain – such as, for example, a stroke.

In the course of our conversation, it turned out that Wahid's wife had died several years ago. He now lived alone, independently. He had three children, but they were all grown up and now living with their own families. They came to visit him from time to time. He'd briefly mentioned to them that he had woken up to see people surrounding his bed, but they had brushed these 'visions' off as some kind of strange dream. Wahid didn't have any hobbies, but in his spare time he helped with a local Muslim charity. He'd been very involved with packing tents and sleeping bags for the growing refugee crisis in Syria. More than two and half years on from the beginning of that country's civil war, hundreds of thousands of displaced people were seeking asylum first in Turkey and Jordan, but now also beyond the Middle East; many were risking their lives in small boat crossings to reach Greece via the Mediterranean Sea.

'Apart from seeing the people and animals that you know aren't there, do you ever have any other kinds of odd experiences?'

'Not really.' Wahid hesitated again. 'But a few times I have thought there is someone else in the room to the side or behind me, but as soon as I look there is nobody there. I am not sure that it is anything important. I don't find that frightening.'

Wahid was describing what neurologists term a 'presence phenomenon' – the erroneous feeling that another person is in close proximity. Also sometimes referred to as an extracampine hallucination (a perception of something outside the perceptual limits of the sensory visual field),[7] these type of experiences often provide an important clue as to what the underlying diagnosis might be.

Next, I performed the neurological examination. I watched Wahid walking, outside the consultation room and down the corridor. His stride length was normal, but it was quite evident that he was not swinging his arms as most people would. Otherwise, his gait and ability to turn was unremarkable. When I asked him to stand still and then gave him a pull from behind, he was able to maintain his balance without difficulty. All the rest of the neurological examination, including the reflexes in his arms and legs was also normal, except for two small features.

First, when I asked him to tap his index finger against his

thumb repetitively as fast as he could, he was slow and the amplitude of the movements reduced erratically within a few taps; this was worse in his left hand. Second, when I moved them, there was some mild rigidity at his wrists; again, it was worse on the left side. The reduction in arm swing when he walked, the slight abnormality of his ability to make tapping movements with his finger and thumb, and the mild rigidity at the wrists were all features of Parkinsonism – characteristic signs observed in Parkinson's patients, but in Wahid's case they were not sufficiently severe to make a diagnosis of Parkinson's disease itself. Nevertheless, they were a potentially important clue as to what might be causing Wahid's visual and extracampine hallucinations.

People with Parkinson's disease have slowness of movement, rigidity and tremor. They lose the swing of their arms when they walk, they may develop poor balance and a stooped posture. However, we now appreciate that there are several neurological conditions which have some of these Parkinsonian features, but which are not Parkinson's disease. I began to wonder if Wahid might have one of these.

We performed some screening cognitive tests of Wahid's ability to attend to information, his memory – both short- and long-term – as well as tests of language, visuospatial ability (such as copying a drawing), and limb praxis (copying of hand gestures). His only difficulty was with the two kinds of copying. He found both challenging. His copy of a line drawing of a cube, for example, appeared haphazard and inaccurate. I checked his eyes and his vision, but these were both normal.

'So, what do you think, doctor? Will they lock me up?' he asked slowly and with trepidation.

'No, I don't think you are mad. These visual hallucinations certainly don't mean that you are mad.'

'Really?'

'Yes, really,' I assured him, 'but we need to do some special scans to look at your brain and then we can discuss whether we can help you.'

'You might have some treatment?'

'Yes, I really think so, but we need to do the tests first. I'll see you back here as soon as I get the results.'

~ o ~

As I left the clinic that afternoon, the conversation with Wahid about madness kept recurring in my mind. He had thought that his visual hallucinations were a sign of insanity and that he might be incarcerated because of this. Psychiatrists make a distinction between hallucinations (false perceptions) and delusions (false beliefs), but in some conditions both symptoms can exist. Many people with schizophrenia, for example, suffer from different types of hallucination (often auditory: hearing voices), as well as delusions (e.g. that they are under the control of an external force or person). However, even healthy people can experience hallucinations under certain circumstances.

For example, as many as half of individuals who have recently been bereaved report hallucinations, while other surveys have revealed that up to ten per cent of otherwise normal people can experience minor hallucinations on a regular basis. They might hear their phone ringing when it isn't, or footsteps coming down a hall when there are none. Why might someone perceive something in the absence of a sensory stimulus? To answer this question, it is important to appreciate that perception is not simply equivalent to a passive 'receipt' of information gathered by our senses. Instead, most neuroscientists consider perception to require an active process of 'reconstructing' the world around us – to *re*present it within the brain.

The nineteenth-century German physicist and psychologist Hermann von Helmholtz, who made major contributions in several fields – including that of thermodynamics – was fixated on a key problem in visual perception: how do we perceive what we see though our eyes? To Helmholtz, the answer was not obvious. How does the pattern of light falling on the two-dimensional retina, he asked himself, become converted into a rich, three-dimensional representation of the world? This was all the more perplexing because he understood that very similar patterns of stimulation on the retina might be caused by different objects, depending on their orientation and location in space. By its very nature, he correctly surmised, the retinal 'image' is ambiguous. Perception is further complicated by the fact that an object looks very different from various viewpoints. Consider a chair seen from the top, or the side, or the bottom. Despite the pattern of light falling on the retina being very different in these three cases, we nevertheless perceive the same chair.

To explain how we solve these challenges in perceiving the world around us, Helmholtz proposed that our brains effectively decide on the most likely solution for which object or objects might be causing the visual sensory impressions on the retina. This type of judgement, Helmholtz argued, occurs by a process he called unconscious inference.[8] By this he meant that our perceptions are actually the result of unconscious assumptions that we automatically make about the world around us. For Helmholtz, perception is not the passive receipt of sensory information, but rather it involves solving the problem of what caused a particular pattern of sensory input by using the observer's knowledge of the environment.

Regularities in the external world mean that we can make use of this knowledge to generate the best interpretation of the sensory information around us. For example, the fact that there are more vertically and horizontally orientated objects constrains what kind of object might be out there. Or the fact that light comes from above to illuminate the environment means we interpret shadows differently for indented objects compared to protruding ones. Some modern-day neuroscientists, who are involved in developing models of the brain, have been inspired by Helmholtz's proposal of unconscious inference. They suggest that the brain solves the problem of sensory input by using a 'generative model' of the external world. According to this view, the job of the perceptual system is to act as a sort of statistical inference engine – a device these neuroscientists aptly named a 'Helmholtz machine'. This 'machine' functions to infer the most likely causes of sensory input, given what we know about the environment we are in.[9]

According to this view, what we perceive is partly based on prior expectation and partly on more immediate sensory input. But this also means that perception can be wrong if the prior probability of something occurring is erroneously taken to be very high. For example, in an experiment where people were told that at some time they are going to hear a particular song, e.g. 'White Christmas', but are instead played white noise, some of them reported that they could hear Bing Crosby's voice. Intriguingly, schizophrenic patients who hear voices are more susceptible to drawing such erroneous conclusions than those who don't.[10]

These considerations have led to the astonishing conclusion that normal visual perception is not like taking a photograph, but is instead an active process in which we generate a model of the world outside. Thus, what we see is mostly about our expectations of what we should see. This means that one way to think about hallucinations is that they are simply *erroneous inferences* about the environment. People with degraded vision, such as that caused by eye diseases, may suffer from visual hallucinations, a condition known as Charles Bonnet syndrome. This is because they are making incorrect models of the world on the basis of poor or 'noisy' sensory input. But we wouldn't consider them to be mad simply because they are experiencing such hallucinations.

Similarly, even if the eyes themselves are working fine, but the parts of the brain that analyse visual information are not functioning properly, then the sensory input may not be effectively processed. And if prior expectations are strong in the face of an individual's imprecise processing of sensory information, then erroneous inferences may be made about the external environment, which can likewise lead to visual hallucinations.[10]

Neuroscientists sometimes refer to incoming sensory information or the processing of that input by the brain as 'bottom-up' processes. By contrast, prior expectations based on previous knowledge or memory are called 'top-down' processes. Visual hallucinations can occur when top-down influences – or prior expectations – are strong but bottom-up ones are degraded. Set in this context, Wahid's experience of visitors in the night was an important clue. His hallucinations seemed worse when it was dark, when the sensory input would be most degraded. But his eyes and vision seemed normal, so we couldn't attribute his hallucinations to an eye disorder.

In his case, the explanation for his symptoms would therefore be that his brain was having trouble processing the visual input. I certainly didn't have to conclude that Wahid was mad because he was experiencing these visual hallucinations. Nor did we have to worry that he was in danger of being placed in an institution because of these symptoms.

~ o ~

The MRI scan of Wahid's brain was performed a few weeks later. Its results did not reveal a diagnosis. However, I had also requested a different type of investigation called a DaT (dopamine transporter) scan, which is a way to assess dopamine-containing

132

neurons within the brain. The outcome of this scan was definitely abnormal. It showed that Wahid had lost some of the dopamine neurons in his basal ganglia in a pattern that is characteristic of two closely related brain diseases. One is Parkinson's disease and the other is dementia with Lewy bodies (DLB). After Alzheimer's disease and vascular dementia, DLB is one of the commonest causes of cognitive impairment in older people.

Both Parkinson's and DLB are associated with the accumulation of Lewy bodies within neurons. These are abnormal aggregates of protein that have clumped together inside nerve cells. They can be seen through a microscope and were first observed by Fritz Lewy in 1910 when he was studying the brains of patients who had suffered from Parkinson's disease, or 'Paralysis agitans' as it was then known.[11] Remarkably, Lewy was working in Alzheimer's laboratory in Munich when he made his discovery. Thus, two very important pathologies underlying neurodegenerative diseases – in Parkinson's disease, DLB and Alzheimer's disease – were discovered within the same room by researchers in the same group, within five years of each other.

We now know that Lewy bodies contain a protein called alpha-synuclein. Whereas amyloid and tau proteins accumulate in the brains of people with Alzheimer's disease, alpha-synuclein aggregation in the form of Lewy bodies is the hallmark pathology of Parkinson's disease and of DLB.[12] In both of these conditions there can be visual hallucinations, fluctuating attention and some degree of cognitive impairment, including difficulties with copying line drawings (constructional apraxia). While Parkinson's disease is associated with motor symptoms such as slowness of movement, rigidity and tremor, none of these have to be present to make a diagnosis of DLB. Nevertheless, people with DLB usually have some mild features of Parkinsonism, and these may become more prominent over time. Many neurologists consider Parkinson's disease and DLB to be two ends of a spectrum of conditions associated with the accumulation of Lewy bodies in the brain.[13]

When I had examined him, Wahid had some clinical signs consistent with very mild Parkinsonism (reduced arm swing when he walked, some slowing of finger movements and a little rigidity at his wrists). These features, combined with his visual hallucinations and mild cognitive impairments, all pointed to a clinical diagnosis of DLB. We now had an explanation for Wahid's symptoms.

~ o ~

A few weeks later, Wahid returned to the clinic. He seemed very different. His face was hollowed, the skin around his eyes now intensely dark, and his gaze was directed to the floor.

'How are you?'

'Not very well, doctor. I have seen more of these people, mostly at night. I'm not sleeping well. My children are very upset,' he said slowly.

'I'm very sorry to hear that. What has happened with your children? I thought you said they hadn't been too concerned.'

'They've now been with me on some evenings when I have had these hallucinations. At first, they didn't believe that I was seeing anything, but then they saw how disturbed I looked. They know that I am having scans and when I got your letter saying that the MRI had been normal, I showed it to them. I could see on their faces that they think this means I must be going mad.'

'You might be jumping to conclusions, perhaps?'

'I don't think so. Normally, they would take me along to some community events or parties. They haven't invited me to anything in the last few weeks. I think they would be ashamed of me.'

'Look, Mr Razzaq,' I assured him, 'you are not going mad. I think we have an explanation for the hallucinations.'

'You do?' There was a curious, hesitant excitement in his voice.

'Yes, the second scan you had — the DaT scan — was not normal. It tells us that you might have a condition called Lewy body disease.' I avoided using the term dementia with Lewy bodies because I knew this would generate more anxiety and, in truth, although the condition was given that name, Wahid wasn't demented. He had some mild cognitive changes, but he did not reach the clinical criterion for dementia which would require his cognitive impairment to be severe enough to cause problems with everyday life. He was living independently.

'I have never heard of this Lewy body thing. What does it mean?'

I explained a little about the condition and focused on the treatment.

'The good news is that we can give you some medication which might help with the hallucinations.'

'Really? You think it might get rid of them?'

'I can't guarantee that, but we can certainly try.'

I gave him a prescription for rivastigmine, a drug that boosts the levels of acetylcholine in the brain. Like dopamine, this is a neurotransmitter used by nerves to signal to each other. Research on people with DLB has revealed that, in addition to the loss of dopamine, there can be a severe depletion of acetyl-choline in their brains.[14] Elevating the levels of this chemical in the brain, therefore, may improve communication between neurons — and, consequently, with sensory perceptions. We now know that those patients with DLB who experience visual hallucinations rely more on their prior expectations than what is actually present in the visual scene.[15] Brain imaging has also shown that the connections between cortical areas involved in processing visual information, especially about objects, are not as strong in these people.[16]

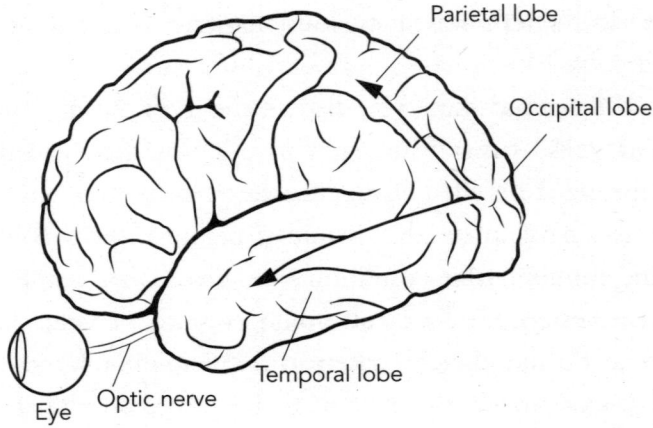

Fig. 11 | Brain regions involved in processing visual information. Information from the eyes is transmitted via the optic nerve to the occipital lobe. From there it passes along two major cortical systems. One of these systems (top arrow) consists of a set of connections up to the parietal cortex. This is important for knowing where objects are around us. The other system (bottom arrow) takes visual information to the temporal lobe, which is responsible for object perception. When connections between cortical regions in these two streams of information are degraded, visual processing of inputs from the eyes is weaker. For people suffering from such degraded bottom-up processes, sensory evidence is even worse when the visual environment is dark. They may rely more on their top-down prior expectations of what should be in the visual scene, but their inferences can be wrong, leading to visual hallucinations.

Visual input from the eyes first reaches the cortex of the occipital lobe in the back of the brain (**Fig. 11**). From there, visual information is processed in two major cortical systems: one going up to the parietal cortex (important for visuospatial functions enabling perception of the locations of where things are around us), and the other going down to the temporal lobe (crucial for the perception of objects). When connections between cortical regions in these two major systems are degraded, as occurs in DLB, processing of incoming visual information from the eyes is weaker. People with DLB and hallucinations are likely, therefore, to rely more on top-down processing of prior expectations and less on bottom-up processing of incoming visual information. By boosting acetylcholine, it might be possible to strengthen the connections between brain regions, including those involved in processing visual information.

'We will start at the low dose, and see how things are going in a few weeks' time.'

'Thank you,' he responded, a smile crossing his face.

'We also have to discuss work and driving.'

Wahid looked alarmed. 'I have to stop working?'

'Not necessarily, but I think you have to discuss the diagnosis with your occupational health team who will contact me. You will also have to let the driving authority and your insurance company know.'

'But they will take away my licence.' He frowned, now clearly very upset.

'No, that doesn't have to happen, because your symptoms are not so severe, but you might have to take a driving assessment, and if the medication is effective that might improve your condition. The problem is that if you don't inform people, there might be serious consequences.'

'Oh no, I will lose everything if I can't drive, doctor.'

'As I say, you won't necessarily have to stop driving and I will explain what I have found to support you.'

Wahid hesitated. 'You really don't think I'm going mad, then? Please be honest, doctor. I am very worried that I'll be locked up in a madhouse.'

'I certainly don't think that will happen, Mr Razzaq, and I honestly don't think you are mad,' I said, trying to be as definitive and reassuring as possible.

There was a long pause as Wahid considered my words. 'Thank you. I really appreciate all you have done.'

Wahid left the consulting room, concerned but also holding out hope for a solution.

~ o ~

The door shut behind him and I was about to start my letter to Wahid's GP. But something in the way that he had expressed his distress over being put away in 'a madhouse' made me stop. Wahid's concerns disconcertingly brought back memories of a surprise that I had encountered several years ago on a historical tour of the City of London, the financial heart of the capital. It was on this tour that I first came across the site of Bethlem, which for centuries had been London's notorious hospital for the insane, the biggest 'madhouse' the city had ever built.

The tour guide's account had embedded into my own mind some shocking imagery of how Londoners had dealt with people who they had considered to be mad, and what the causes of their insanity were imagined to be. Originally founded in 1247, Bethlehem Hospital – soon shortened to Bethlem – had first been located in Bishopsgate, we were told. As early as the fourteenth century it was a place known for housing the mad, the outcasts of London. It was where people who didn't belong, sometimes simply because their behaviour didn't conform to society's expectations, were interned. Bethlem was soon being referred to as 'Bedlam', a word synonymous with the pandemonium inside its walls. Because the burgeoning need to house 'the lunaticks' had outgrown its original confines, the hospital moved in 1676 into an imposing new building in Moorfields by London Wall, a street that marks the location of the old city wall. It was here that our guide had stopped to point out its location, just by present-day Finsbury Gardens.

In Georgian times, she explained, the governors of Bethlem actively encouraged visitors to come 'view' the patients, even if they had no connection to the institution's inhabitants. Presenting the misfortunate in this way was an important means by which Bethlem elicited donations – a fundraiser of sorts. The spectacle could also be framed as moral instruction for the public: the disturbed folk housed in Bethlem were a cautionary reminder of where vice and degenerate behaviour might lead. But, as I later read, what also drew people to gawk at the poor inmates 'was the *frisson* of the freakshow'.[17] Here the Georgian public witnessed all that there was to offer in 'Bedlam': the chaos and vulgarity of the mad as a spectacle. This was 'othering' on a showground scale.

An example of what they might have encountered, our guide went on to say, is depicted in the last scene of *A Rake's Progress*, a satirical series of paintings by William Hogarth. The story told across these portrayals is of the life of the reprobate Tom Rakewell, a man whose depraved lifestyle – gambling, prostitution and debauchery – ultimately leads to ruin, culminating first in his internment in the squalor of the Fleet Prison, used mainly for debtors and bankrupts, and then confinement at Bethlem. In this last scene, set in a filthy cell in the hospital, he is hardly recognisable from his previous dandy form. Half naked, his head shaven and his leg manacled, his face is in a contorted grimace, and around him are other examples of his fellow inmates, including rather presciently the figure of a king, naked and wearing a crown on his head as he urinates in his cell.

As our tour guide explained, the numbers attending Bethlem to ogle at its inhabitants were particularly large on the holidays at Easter, Whitsun and Christmas. It was almost like a visit to the zoo. The crowds would jostle to get the best sightings, shouting, cajoling and even prodding the afflicted. The period of 'viewings' reached its peak under the directorship of the hospital by John Monro, one of a family dynasty of 'mad doctors' (the name given to physicians tending to its residents) who oversaw Bethlem for 125 years.[18] When in 1788, King George III became seriously unwell and many thought he was incurably insane, the royal doctors wrote to Monro for his opinion. He proved to be wholly indecisive on the matter.[18] Instead, Monro's son Thomas was requested to attend. King George – paradoxically, rather sensibly – would have none of it, and proved to be extremely reluctant to be treated by a doctor from the infamous Bethlem.

By then, though, the King's behaviour had deteriorated dramatically. At times he was extremely agitated, speaking rapidly almost as if delirious, and difficult to reason with. On other occasions, although he seemed calm, he would soon begin to express delusional thoughts, convinced, for example, that London had been flooded and that he was the cause. He wrote bizarre letters and insisted that honours should be bestowed on anyone who approached him.[19] He hardly slept and also experienced visual hallucinations and misperceptions. On one occasion, for example, he was observed to be trying to shake hands with an oak tree while talking to it, seemingly believing it to be the king of Prussia. However, the royal doctors could not agree among themselves on whether he was insane, let alone if he was incurable.[20]

Was the King's behaviour sufficient to conclude that he was now mad for the rest of his life, or had it been triggered by some physical illness that would pass, or might even be treatable? Hogarth's satirical engraving of physicians ruminating over a puzzling case entitled *The Company of Undertakers* captures the widespread sentiment of the time: that much of Georgian medicine was performed by medical experts who were no better than quacks. As the King's illness continued without showing any signs of abating, the pressure on the royal physicians increased daily both from the King's household, including Queen Charlotte, and from the government whose political fate was tied to that of the survival of the monarch. Should the King be pronounced permanently insane, thereby triggering the coronation of his son, the Prince Regent, or was there hope of recovery? Nobody knew.

Ultimately, it was decided to bring in the services of a physician not associated with Bethlem, Francis Willis, under whose care, as we learned in Chapter One, the King did get better. His improvement triggered further pronouncements by different doctors that he couldn't possibly be insane. This must have been a physical illness, they declared. And, indeed, the question of whether mental illness has an underlying physical cause has a long history. Modern biological views of psychiatric illnesses, such as schizophrenia or bipolar disorder, would argue that all of these must ultimately have a physical – neural – basis.[21] How could they not? So, the distinction between physical and mental illnesses is in some ways artificial. But this is to take a very deterministic view.

Others, like Michel Foucault, philosopher and historian, have radically reframed 'madness' as a construct imposed, over the ages, by society on people who simply behave differently to the rest of us.[22] In Foucault's view, people who don't fit are the ones who have been labelled as insane. He traces the history of madness from medieval times through to what he terms the 'Great Confinement', the development in seventeenth-century Europe of incarceration of 'the insane' as a means to segregate these people who, just like vagrants, blasphemers or prostitutes, were best kept separate. In Foucault's view, Christian societies considered all these groups to be in moral error for having chosen their way of life. They needed to be cast out but also given an opportunity to reform in asylums – the great hospitals that sprang up across Europe at that time.

This is how the modern 'madhouses' started and our views of them have not been helped by more recent representations in films such as *One Flew Over the Cuckoo's Nest*. I could understand how Wahid was very apprehensive over the prospect of any form of 'confinement' because of his symptoms. Who wouldn't be?

~ o ~

A few weeks later, Wahid returned to the clinic, but this time with a smartly dressed young woman who bore a striking resemblance to him. With very high set cheekbones, framed by straight, lustrous black hair, her pale face had a peculiarly angular haughtiness to it. She scrutinised me intently through tawny brown eyes.

'This is my eldest daughter, Yasmin. She is an accountant,' he explained, broadening his lips with a hint of pride.

'Good to meet you, and how have you been, Mr Razzaq?'

He didn't get a chance to reply.

'He is not at all well. Are you sure you have the right diagnosis, doctor?' she asked curtly. This was not going to be an easy consultation.

'What's been happening, Mr Razzaq?'

Yasmin answered for him again. 'He is still seeing these awful things at night. There are people around his bed, mice running over the floor. The medicine you gave him hasn't done a thing,' she said, her voice rising as she spoke.

Her father looked despondent.

'His work has put him on sick leave and he's not eating properly,' Yasmin continued. 'Are you really sure he has this Lewy body disease?' Yasmin was clearly not convinced. I tried to slow the pace of the conversation and give Wahid a chance to speak.

'I'm really very sorry to hear that. Mr Razzaq, tell me what's been happening.'

'Yasmin is right. I am really no better, doctor, and they have asked me to go on sick leave for the time being. I've tried to get more involved with the charity I work for, but even they have asked me not to come when they realised that I was having hallucinations. I feel useless,' he concluded in a low voice, utterly dejected. Yasmin looked angry.

'Is there nothing better you can offer him?' she asked. 'We're desperate. My father is losing all his self-respect. He is no longer the man he was. People don't see him. Our family friends in the community aren't calling. You know how superstitious they are about things like hallucinations. Dad is very isolated. I have had to bring him to our house. He's now living with me and my family because I'm worried about the way he is.'

'I understand this is very difficult for all of you. We have only started the medication at the lowest dose. Have you experienced any side effects, Mr Razzaq?'

'No, I don't have any side effects, but it isn't doing anything either.'

'A medical friend of mine was surprised that he is not on an antipsychotic drug for these hallucinations,' added Yasmin. 'Wouldn't that be better?'

'I am more than ninety-five per cent confident that the diagnosis is Lewy body disease,' I said. 'Antipsychotics can actually make people with this condition dangerously ill, so we avoid those if we possibly can. The medication I have given your father does help most patients but not everyone, and certainly not always at the low dose, so we need to give it a chance and increase the dose slowly.'

'So, you're not one hundred per cent sure it is this Lewy body thing,' Yasmin snapped.

It is difficult to explain to patients and their families that with brain disorders it is often impossible to be absolutely certain of the diagnosis.

'To be that confident we would need to look at a sample of your father's brain under the microscope to see if he had Lewy bodies, and the only way we could do that would be with a biopsy of the brain. That's a procedure that can be risky and is reserved only for cases where the diagnosis is not at all clear from the clinical assessments and investigations. In your father's case, the visual hallucinations, neurological examination and the DaT scan result all make it extremely likely that this is the diagnosis.'

'But not certain?' persisted Yasmin.

'That's right, not certain.'

'Could we have a second opinion, then?' Yasmin continued.

Her father looked at her, embarrassed. In his eyes, she was challenging my authority.

'Yasmin, don't speak to the doctor like this,' he chided.

'What?' There's nothing wrong in asking for someone else's view, Abba,' Yasmin responded quickly.

'No, there isn't any problem with that. I can write to a colleague to arrange it. But, meanwhile, I would like you to try a higher dose of the rivastigmine medication. Would you be willing to do that?' I asked Wahid. 'The only way we can find out whether the tablets will help is to slowly go up in dose.'

Wahid nodded his head. 'Yes, I will try.'

Yasmin frowned with scepticism. 'How long do we have to try for?'

'Please bear with me. Sometimes it can take weeks or even months, but we have to give it a go to see if your father can benefit. If we stopped it now, we would never know if he might have improved.'

Yasmin's expression made it clear that she didn't have much confidence.

'Is there no other treatment? My friend told me about deep brain stimulation. Why isn't my dad being offered that?'

I couldn't help raising my eyebrows. Yasmin was referring to the neurosurgical technique of inserting electrodes deep within the brain, into the basal ganglia. It had been shown that electrical stimulation of these electrodes could improve movements in some people with Parkinson's disease. Patients became faster and less rigid. However, there wasn't any convincing evidence at that time that such a procedure could treat visual hallucinations. It was nevertheless a remarkable new advance in treatment for brain disorders and has since been used in a research setting to investigate whether it might improve some of the non-motor symptoms in both Lewy body disease and Parkinson's disease.

'I'm afraid that deep brain stimulation is not being used to treat hallucinations,' I explained. 'Even if it were, this would take months of assessment and planning. It's important that we try to treat your dad right now, so let's increase the dose of the drug and I'll see you again in a few weeks.'

~ o ~

I really hoped that the higher dose of the rivastigmine would make a difference for Wahid. He was so very distressed, as evidently were his family. The words of his daughter stuck particularly in my mind that evening. As I tell our medical students: we carry our patients' problems home with us. Yasmin had been very frustrated, but it wasn't her request for a second opinion that was bothering me. She had pointed to something else that was important: her father was no longer the man he used to be, and people didn't want to see him because they were 'superstitious' about his symptoms. The experience of suffering from visual hallucinations was seemingly sufficient to trigger his increasing isolation. The things that mattered to him in his daily life – his job, his voluntary work for a charity, and his links with his community – had dissolved over a few months. Maybe they hadn't gone for ever, but for the time being at least they had disappeared. Why?

The work concerns were understandable. The bus company needed time to obtain medical reports and driving assessments. Unless there was some resolution of his symptoms or a plan for the future, his diagnosis was always going to challenge his ability to continue to work as a bus driver. Perhaps the same might also be said for the charity's response. But the loss of connections with members of his own community couldn't be attributed to that. One of the most impressive qualities of immigrant groups

145

from South Asia living in the UK has been the mutual support people give to each other in difficult times, including when members of their community fall ill. Families rally around, offer to help look after the children, cook meals or are simply able to be there to comfort the sufferer.

Although there are now third- or fourth-generation migrants from Pakistan, these communities often still remain outsiders in many ways. For many, acculturation – the process of learning and assimilating the values and customs of the host country, including behaviours that affect health – has not been substantial. As John Berry, a Canadian social psychologist who has studied the trajectories that migrants take, has observed, while some groups integrate, other rarely interact with the host community.[23] Instead they stay separate, within the boundaries of their own social group. In so doing, they need to stick together to confront challenges such as ill health within members of their community. But what had happened to lead to Wahid's near abandonment, and sense of exclusion, when he became unwell?

He hadn't announced he was sick, but Wahid's responses to his visual hallucinations had been observed by people in his community. They soon realised the nature of his symptoms, and had made judgements about what that meant. It wasn't necessarily the case that they thought he was possessed, but the knowledge that he was experiencing hallucinations was sufficient for them to shun him.

~ o ~

A month later, as I looked down my clinic list to see who was coming that day, I saw that Wahid was due back. I was dreading it. If he hadn't improved, there was likely to be another very awkward conversation with his daughter. I would have to hold my ground and try to convince them both that we needed to increase the dose of rivastigmine even further. Persuading them to take this route would mean a far more difficult conversation than the one we'd had previously.

As anticipated, Wahid came with his daughter, Yasmin. Her mouth was already half open to speak before she had even got in through the door, and her eyes immediately fixed me with a grave intensity. But it was her father this time who beat her to opening the conversation before I had welcomed him in.

'Doctor, it has been marvellous. The higher dose has worked! I am so very pleased. No more of these terrible things coming to visit me in the night!'

I was taken aback.

'That is fantastic, Mr Razzaq, quite wonderful.'

But before I had any opportunity to enjoy the moment with her father, Yasmin interjected and, to my surprise, she was grudgingly appreciative.

'I'm sorry for being so difficult the last time,' she explained. 'You can understand that I was just worried about Dad and, frankly, I didn't really expect much from this medication.' With an expression of slight self-reproach, she went on. 'I was wrong. Thank you for explaining things clearly. I should have trusted you. He really is a very different man and I know that it is down to your help.'

This was not what I had expected, but it was very gratifying.

'I understand that you were stressed. I am so pleased your father has responded well,' is all I could say.

We often don't know whether a drug will work for any given individual patient. To learn that rivastigmine had made such a difference was really pleasing to hear. It had apparently taken a week or so before the higher dose of the drug had its effect, but now Wahid was completely free of visual hallucinations. It is in some ways remarkable that a drug that boosts the level of a specific chemical in the brain – the neurotransmitter acetylcholine – can achieve this. As I was to find out, it had effectively turned someone who might well have been considered mad in years gone by – or even in some communities today – into a person who was 'acceptable'.

We now know that many patients with Wahid's condition can obtain substantial benefits, including improvements in cognitive functions and resolution of visual hallucinations, with rivastigmine and other drugs that increase acetylcholine in the brain.[24] Intriguingly, the major source of neurons that use acetylcholine as their neurotransmitter emanate from a tiny brain region called the nucleus basalis of Meynert (or NBM for short). One function of acetylcholine is to amplify signals transmitted across neurons. Rivastigmine might strengthen the connectivity between cortical areas involved in processing incoming visual information, which are weak in Lewy body disease.[16] In this way, it may help to enhance visual perception, and thereby also reduce the reliance on prior expectations in these patients.[15] In recent years, neurologists have begun to research whether electrical stimulation of the NBM can also improve cognitive and neuropsychiatric symptoms such as visual hallucinations. These pioneering deep brain stimulation studies have shown that, at least in some patients, there can be significant benefits.[25,26] The same technique is also now being tested in Alzheimer's disease.

Buoyed by the improvement in his symptoms, Wahid had summoned up sufficient courage to revisit his occupational health team. They had said that they'd like to see whether he remained free of hallucinations over the next two weeks. They also wanted a report from me before making a decision on whether he might return in some capacity to work. In the meantime, without prompting, Wahid had arranged a driving assessment, which he had passed without any difficulty. Most pleasingly for him, he had been offered a place again as a volunteer back at the charity that was so dear to him.

Yasmin had also made a big effort to contact his family friends.

She had reassured them that her father was much better now on his appropriate medical treatment. To reduce the stigma associated with his symptoms, she had apparently had to emphasise that he had been seen by a medic: a neurologist, not a psychiatrist. Three weeks ago, Wahid had moved back to his own house. Yasmin felt that he was now far less vulnerable, and it was important for him, she felt, to regain his independence. Some of his friends had started to call around and he was enjoying their company, even being invited to community events once they had seen that he appeared to be back to his normal self. Wahid's story wasn't over, but the beginning had changed. There were no more visitors in the night.

5

Neglecting me quietly

The light was fading fast over Portobello. It had been a grand evening. The deep violet of the west London sky was scratched with a lush ruby red at its lowest edge as the sun slowly waned on the horizon, announcing the end of a long, scalding summer's day, the kind that had rarely frequented the city that year. People were making the most if it, and where better than in Notting Hill. The pubs were overflowing with customers stacked two or three deep on the pavement, most in a radiant mood, revived by both the heat of the day and London's finest ales. Their good humour and laughter stretched far along the streets.

It was then, apparently, that Winston had been found. At first sight, he had simply been having too much of a good time. Twisting and turning in his old pinstriped charcoal suit, onlookers would have assumed that he was probably a little worse for wear, attempting to dance to the beat of the distant reggae tracks of Burning Spear booming from somewhere near the Westway. He was shouting something, but then so were many others. It wasn't clear what he was saying, his Caribbean lilt distorted by a slight slurring of his speech. No doubt this might also be attributed to the cumulative effect of the libations enjoyed throughout the day.

When his friend Kelvin bumped into him, Winston didn't seem to see him at first. He appeared somehow bewildered, peculiarly dazed in the growing darkness. Although Kelvin had crossed over the street, greeting him as he approached, Winston's gaze was firmly turned away from him, attempting to find the owner of the voice which had called out his name. Even when Kelvin was straight in front of him, Winston didn't appear to see him, turning rightwards away from him in an attempt to locate him.

'Is that you, Kelvin man?' Winston had asked.

He seemed disorientated, unsure of where he was even though this was his neighbourhood. Kelvin was concerned.

'You OK, Winston?' he had asked. 'Been drinkin' already?'

But Winston was adamant that he hadn't, and there certainly wasn't any smell of alcohol on him.

'Me need a piss, though,' he said.

Kelvin had put his arm through Winston's and walked with him to the nearest pub, The Elgin, at the corner of Ladbroke Grove. The place was well known to them both so, once inside, he'd let Winston go ahead towards the toilets while he went to the bar. But just as he was about to order two pints, he heard a terrible crash. As he turned, he saw that Winston had apparently collided with a woman carrying some glasses. Both were now entangled on the floor with people trying to help them up.

'Jesus, Winston!'

Kelvin had rushed over, apologising for his friend as he slowly pulled him up. It became evident, though, that Winston must have been heading in the wrong direction, back towards the front doors rather than the toilets.

Once the commotion had died down, he decided to guide Winston to the toilets himself. He was now even more concerned that all was not well. They both went into the lavatories and Winston headed over to the urinals, but to Kelvin's horror, he saw his friend unzipping and pissing straight into the sink. The other men there shouted that he should take his drunk friend away. Hadn't he caused enough trouble already? Although Kelvin had insisted that Winston wasn't drunk, he decided to heed their advice and shepherded him to the bar. Convinced that Winston must be ill, he asked the owner to call for an ambulance while he sat his friend down in a corner.

'Kelvin man, stop making a fuss. Me alright. Just let me go home,' Winston had apparently complained. But Kelvin wouldn't have any of it and managed to persuade him to stay.

When they arrived at the emergency department of the hospital, they were confronted by a very long queue.

'So, your friend's confused, huh?' the triage nurse had asked, nodding her head with disapproval, peering down her spectacles with obvious scepticism. She'd seen many 'confused' patients that evening and didn't believe Kelvin when he protested that no alcohol was involved, and so they waited to be seen by a doctor. When eventually the moment arrived, some five hours later, nearer dawn than dusk, the young man who assessed Winston decided to send off some blood tests to see if there might be a reason for his delirium, for that is what he concluded must be the diagnosis. Winston, as previously, seemed to be somehow dislocated from his surroundings, not being able to cooperate with the physical examination, always turning his head, as if distracted and looking for something.

'I'm afraid it'll take a couple of hours before we get some of these blood tests back,' the young doctor said. 'If, as you say, he hasn't taken anything, it's possible he has an infection. His chest sounds clear, though. We'll get the nurse to take a sample of his urine. The blood tests may also help us work out if he's caught a bug.'

So, they waited further, Winston falling asleep on his friend's patient shoulder.

The sights and sounds of the emergency department at that time of night can seem like a circus. Kelvin, unable to doze off, might have observed some of the regulars that pass through its doors overnight. There might have been Will, the musician who always wore a bright orange suit, wheeling in his double bass confidently past the receptionist, seemingly perfectly well but explaining to all around him that he was in pain and needed his 'oxy' (oxycodeine), if he was to make it through the night. Or else young Jade who, intent on avoiding the attentions of several dubious men, had arrived once again with complaints of abdominal pain which never did materialise into a firm diagnosis. Or homeless Karen with her shopping trolley full of her worldly possessions, feeling unsafe to be out on an evening when so many people were drunk, even if it was warm. All the richness of humanity passed through the emergency department at night.

~ o ~

That morning was my turn to do hospital rounds with the junior neurologist, Nicky, who was on call. I'd got to know her only over the last couple of weeks. My first impression had been that she was efficient more than enthusiastic: a doctor who liked to tick jobs off, rather than become engrossed in the problems that people presented with. Even though it was early, she already had a long list of patients referred to us by the medical and surgical teams over the last twenty-four hours for neurological opinions.

'Let's start with the emergency department,' I suggested. 'We can probably help them clear a few of their cases quickly.'

'OK. The first one's a seventy-year-old with acute confusion,' she observed, rolling her eyes upwards with a deliberate look of derision.

'Confusion' is a term often used by doctors, but it is usually not very helpful. It obscures what might be the underlying cause. For a neurologist, the question is really whether there

153

is a generalised process such as an infection or the effects of a drug, electrolyte (salt) or metabolic disturbance that is causing someone to be delirious because of a *global* brain dysfunction, or whether instead there is a distinct, *focal* brain injury that might make someone appear to be confused when they aren't. For example, if a patient had a language impairment, they might sound very confused, when the underlying cause might actually be in comprehension or in speaking.

'Well, let's see him, then,' I agreed.

~ o ~

When we got to the emergency department, Winston was behind the curtain in a small bay, lying on a 'trolley', a euphemism for a skeletal mobile bed. He was clearly unhappy being there, as he explained to Kelvin in very robust terms in his Jamaican patois.

'Why you bring me here, man? Nothing not happenin'. Doctors dem not doing nothing. Understand me?'

'Winston, look! Here are some docs now,' Kelvin said soothingly as we approached.

'Where, I don't see no docs!' Winston retorted, turning to his right.

'Straight in front of you, man. Look over here,' Kelvin tried to point out.

'Yes, we're here to see you, um . . . Mr O'Hara. It is Mr O'Hara, right?' Nicky asked, questioning whether we had got to the right patient.

'That's me alright: Winston O'Hara. You expectin' someone white, doc?' Winston grinned, his head turned to the right. 'Everyone does! My ancestors are Irish Jamaicans, just like Bob Marley's. Bet ya didn't know that? Yes, Marley was Jamaican Irish alright,' he said with a laugh.

'So were Marcus Garvey's ancestors. No?' I asked.

'Well, well. We got a smart one here.' Winston smiled, nodding his head. 'Quite right, doc. Now ya smart enough to figure out why me can't go home?'

He seemed in good spirits despite his long stay in the emergency department. Winston's face was extraordinarily youthful for someone in his seventies. His skin, a radiant light coffee, was hardly lined. The only thing that might give away his age was the white, curly knots of his hair, trimmed short, and a striking contrast to his skin.

We tried to take the history of his admission from Winston but, now getting more impatient and disgruntled, he insisted there was nothing wrong with him. He'd just gone out for the evening to enjoy himself, he said, but somehow things seemed 'peculiar'. He couldn't 'put his finger on it', but they weren't right. But he was sure a proper night's sleep would 'sort it all out'. We got far more from Kelvin. He related to us the details of the night, how he'd found his friend walking bewildered in Notting Hill, how he'd taken him to the pub but was shocked at how oddly he'd behaved in the gents.

'Ya prone to exaggeration, man!' Winston huffed.

Kelvin said he'd have to go home to pick up a few things for his friend but he'd back later. Winston was not amused.

'Ya betta make sure ya do come back ya know, Kelvin! Don't leave me here. Ya 'ear what I say!'

I asked Nicky to examine Winston neurologically while I watched. She was thorough, moving smoothly from examining his visual fields (testing whether he had any loss of vision), through the rest of the cranial nerves (the movements of his eyes, face, tongue and neck, and sensation over his face), to testing his arms and legs (power, sensation and reflexes).

'All seems normal,' she concluded as she finished, shrugging her shoulders. 'I don't think there is a neurological cause for the confusion. Shall we just say "not for us"?'

She was clearly keen to wash her hands of this confused patient, to cross him off our list. It was true that her examination hadn't revealed anything abnormal, but something in Winston's manner, the way he kept turning his head to the right, rang alarm bells.

'Let's just see how Mr O'Hara walks,' I suggested.

Winston fairly jumped off his bed, eager to show us that he was fit to be discharged, and drew the curtains back.

'Me walk just fine,' he said as he headed off. And he did move well. But within a few steps he had veered so much to his right as he headed down the congested corridor that he was bumping into trolleys parked by the wall.

'I can see why they thought he was drunk!' Nicky observed wryly as she studied his attempts to unravel himself from the curtain of one of the other examination bays.

'Don't you think it's interesting that he's always turning to his right, though?' I asked.

We brought Winston back. I sat him down on a chair.

'Can I get you to take a look at this newspaper?' I asked. 'Tell us what's been happening in the world. Are there any interesting stories here?' I asked as I picked up and unfurled in front of him a creased copy of the paper. It was the *Daily Mail*, one of the UK's notorious tabloid newspapers, which happened to have been left by a previous occupant of this examination bay. He'd be lucky to find anything interesting in this rag, I thought to myself, but let's see what he does.

Winston surprised me by engaging with my request.

'Well, looks like the Americans are bombin' those Islamic State fighters in Syria. I really don't understand what that's all about, I gotta say.' He was referring to President Obama's response to the taking of American hostages by Islamic State, more than three years on from the beginning of the war in Syria. Winston turned the page. 'Then there's this newspaper guy goin' to prison

156

for hacking phones of famous people. Serves 'im right! And then two boats full of people escaping from Syria have sunk in the Med, with a lotta people dead. It's so sad.'

He was understanding each of these articles, but what was clear was that he was looking at only the right page of the paper as he turned it, missing out the news items on the left.

'What about this article?' I said, pointing to one on the left-hand side, a report on the effects of the hot weather. Winston now turned towards it and also recounted what it said, but as soon as he finished his gaze shifted rightwards. Every other story he found spontaneously was on the right side of the newspaper as he turned the pages one by one until he got to the end.

'Have you got a sheet of paper?' I asked Nicky.

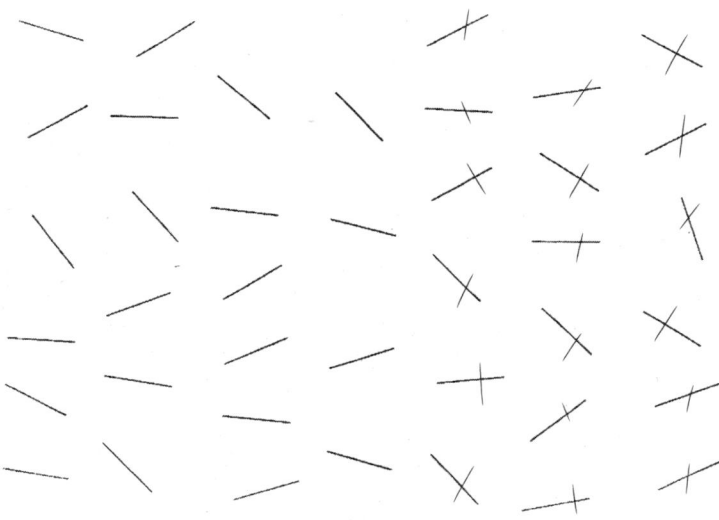

Fig. 12 | Visual search performance. Winston crossed out only lines to the right side of the page, neglecting those on the left.

'Sure,' she replied, handing me one of the medical history sheets she had brought along with her to document our findings. I drew a bunch of short lines across the page, then placed the sheet straight in front of Winston. Offering him my pen, I asked him to cross out all the lines he could see. Unlike most people, he started off crossing lines at the very right of the page. (Most people who read left to right will start on the left.) He stopped after just half a minute, leaving most of the lines on the left side of the page untouched (**Fig. 12**).

'Are you done?' I asked.

'Oh ya, now why you aks me to do this kind of silly thing for?' Wilson replied indignantly, as if I had requested a task that a child could do.

'Thanks for bearing with me,' I responded.

I turned to Nicky. She was looking down at her feet.

'Sure got that one wrong,' she concluded sheepishly. 'He's got left-sided inattention, hasn't he?'

'I think he has,' I responded. 'That explains the veering off and turning to the right. He's not confused at all. He's neglecting things to his left.'

Just at that moment, the young emergency doctor who had referred Winston to us pulled back the curtain to the bay to announce that all the blood tests he'd so far got back were normal.

'That would be the case,' I said.

'So, what do you think is causing his delirium?'

'I don't think he is delirious. Mr O'Hara here has had a stroke.'

'What? But he doesn't have any weakness.'

'He doesn't have to. If the stroke didn't affect parts of the brain that are involved in controlling movement, there won't be any weakness. But he does have quite profound left-sided neglect. He's not paying attention to his left. That's why he can't seem to find you when you're talking to him. And that's why he's always looking to his right.'

'Oh!' He nodded.

'Do you think you might arrange for a CT scan of his brain?

We'll come back later to review the results.' I turned to Winston, as I continued. 'Mr O'Hara, I'm afraid I think you might have suffered a small stroke that's making it difficult for you to be aware of things to your left. We're going to arrange for a scan of your brain to see if that's the case, but I'm afraid it wouldn't be safe for you to go home yet.'

'A stroke?' Winston replied, obviously surprised and also sceptical about my diagnosis. 'You sure, doc?'

'I can't be absolutely sure, of course, so let's wait for the scan.'

I was obviously not impressing him with my diagnostic skills. As we left the emergency department, Nicky said, 'I feel stupid for missing the inattention. It was obvious, wasn't it?'

'Only if you've seen it before,' I reassured her. 'Everything's easy in hindsight. You're unlikely to miss this again because now you'll be looking out for it.'

She smiled. 'Yes, I sure will. I'll look up the CT scan result as soon as it's done.'

'Where do you think the stroke will be?' I enquired, interested to see how well she knew her neurology.

'Right parietal cortex?'

'Yes, I agree. That's the most likely explanation.'

'And it's unlikely to have encroached as far as the motor cortex or the corticospinal tract,' she went on. 'Otherwise, we'd have found him to have been weak in his left arm and leg.'

The motor cortex on the right side of the brain controls movements of the left arm and leg through nerve fibres which run in the corticospinal tract. These transmit motor cortex signals to the spinal cord which, in turn, makes the muscles of our arms and legs move in a coordinated fashion via our peripheral nerves. It is truly a remarkable control system.

What we had observed with Winston, though, was that his control of movement was not impaired. Instead, he was profoundly inattentive towards his left. He didn't spontaneously look that way. Instead, his gaze was shifted to the right. When people approached him from his left, he appeared distracted because he was trying to find them towards his right. When he walked, he neglected items to his left, sometimes bumping into them. He wasn't blind. He could see alright, as Nicky had demonstrated very convincingly when she had examined his vision. It was just that his attention was drawn to the right, as we'd also observed when he was asked to read stories in the newspaper.

~ o ~

The CT scan was performed a few hours later and sure enough it showed a relatively small stroke confined to the right parietal cortex. Many years previously, in the 1940s and 50s, several neurologists had observed and documented the phenomenon of leftward inattention following injury to the right parietal lobe.[1,2] It has since become evident that such spatial neglect[3,4] – sometimes shortened to 'neglect' – is a common disorder following a lesion (damage) to the brain, particularly after a stroke, where a blood vessel may become blocked suddenly by a clot or burst to produce a haemorrhage.[4] Patients with neglect fail to pay attention towards one side of space.

But what is attention? Psychologists and neuroscientists have been intrigued by this question for over a century. Our brains are continuously bombarded by information from the external world. Each millisecond our sensory receptors – for vision, hearing, touch, smell and taste – may be activated by varying inputs. Yet our brains have limited capacity. They simply cannot process all this information. Nor would it be useful for them to do so, because most of the inputs that are continually stimulating our sensory receptors are not very significant for us. For example, the touch of our clothes on our skin surface is continuously present and changes even with the slightest movements, but this is usually not of great relevance for most of our everyday activities. We don't normally care how our shirt is brushing against our skin. If on the other hand, we suddenly noticed that a part of our skin was feeling very hot because we had inadvertently moved it close to the flame on a cooker, that would be important. Similarly for vision. Most of the information that is reaching our eyes every millisecond is not very useful, but some of it is. For example, if we were walking down a street where some brickwork was falling from scaffolding, we might need to react speedily to avoid getting hit. Our brains require some means to select the relevant from the irrelevant – and that process has been termed 'attention'.

Researchers working on how we process sensory information have long appreciated that we would be swamped by a deluge of information overload if we didn't have such a process of selection. In fact, precisely this issue emerged when Google came up with the idea of its Google glasses. These would record a person's activity through the entire day so that, even better than a diary, the entirety of our lives would be available for playback. It soon became apparent, however, that most of what we do during a day isn't that interesting – and certainly not worth devoting terabytes of data to store for posterity. Instead, what we might need to capture are just a few crucial episodes during a day: the events that were most significant for us. But which events are those?

The process of choosing which information is relevant is given the term 'selective attention'. It is the means by which our brains filter out information that is not useful, to focus instead on the stuff which might be. Some researchers have proposed that because the brain has limited capacity, sensory information needs to compete to be selected.[5] The higher the salience of the sensory input (whether it is sudden heat on our skin or a projectile moving towards us seen in the periphery of our vision), the more likely that we might choose to pay attention to it. Salient stimuli win in the competition for selection. But where we decide to deploy our attention also depends upon our goals at the time. If, in a crowded railway station, we are looking for a friend who has told us that she'll be wearing a red coat, we'd be trying to filter out any colour that isn't red and focus on those places in the visual scene that are red. So, attention works by competition, either bottom-up (salience) or top-down (depending upon our goals or what we are seeking to do).[6] The things that win in that competition capture our attention.

Many research experiments that have been performed in visual attention have suggested that it can be highly spatially selective: we can attend to one location in space and ignore others.[7] This has led to the view that visual attention is rather like a 'spotlight' beam. It picks out the location in the scene where there is something that might be important. Modern brain-scanning studies performed while healthy people have to attend to one or other side of space have revealed that this spotlight mechanism might be directed by the parietal cortex, with the right parietal cortex shifting attention to the left side of space and the left parietal cortex shifting it rightwards.[6] When the parietal cortex is damaged on one side, people fail to direct their attention towards the opposite side of space.

In neglect, the visual fields can be intact: patients can see, just as Nicky had found when she examined Winston. But their natural tendency is to pay attention to objects towards one side of space. So, after right parietal brain injury, such as the stroke that Winston had suffered, patients neglect the left side of space, attending to items on their right and ignoring those to their left.[8,9] This situation can be considered as though objects on the right are sharp and highly salient for the brain so that they win in the competition for selection and are attended to. Whereas those on the left don't stand out and therefore don't capture the patient's attention. As a result, people such as Winston are not consciously aware of objects to their left. Even if a patient with neglect understands that they might not be paying attention to one side of space, they may still fail to orientate themselves fully in that direction. This was the evident, for example, in case of the famous film director Federico Fellini, after his right parietal stroke.[10]

A much more severe form of inattention than neglect, which affects both sides of space, was first reported in 1907 by the Hungarian physician, Rezsö Bálint. He was fascinated by a man who first came to see him because he had suddenly become frightened to go out of his house. The patient declared that he was unable to judge where or how far oncoming trams or carriages were from him as he walked out of his home in Budapest. He also complained that he could no longer read a book or newspaper. An initial diagnosis that some doctors had considered was that he was simply suffering from 'hysteria', some form of psychological condition.

Bálint, however, was a meticulous clinician. He noticed that if he placed an object straight in front of his patient the man was aware of it, but that if he showed him two objects, he consistently reported seeing only one of them. Through careful bedside examination, Bálint came to the conclusion that his patient had a severe restriction of attention, which meant that he was unaware of any objects either to his left or right. Even if they were placed straight ahead of him his attention was so limited in capacity that he could consciously perceive only one item at any one time, despite the fact that his sensory visual fields seemed intact. Bálint's patient also had difficulty moving his eyes towards or reaching visual objects. But this might have been because he didn't know where the objects were in space. This is why he was unable to judge where the trams or carriages were and why he was so frightened to leave his home. It also probably explained his difficulty reading, as he was unable to perceive the position of words on a page.

When the man died a few years later, Bálint found out at the post-mortem that he conducted that his patient had suffered two strokes, one on either side of the brain, involving the parietal cortex on both the left and right sides of the brain. This led him to conclude that his profound visual inattention, as well as his poor localisation of objects around him, must be due to damage of the parietal lobes. The features of his celebrated case were subsequently eponymously termed Bálint's syndrome, but hardly anything was known about the mysterious Bálint himself.

In 1986, curious to discover a little more about him, I travelled to Budapest from Oxford, where I was working for my PhD. Hungary was still behind the Iron Curtain at that time, so I was apprehensive about how my trip to a Communist country would unfold, but the Hungarians I met turned out to be very generous, welcoming people. To my surprise, after making a few enquiries, I found that Bálint's grandson, Peter, was then Professor of Physiology at the university. Rather sheepishly I knocked on the door of his house, to be greeted with some degree of bewilderment. He found it difficult to believe that a young brown man claiming to be from England had travelled such a long way to discover more about his grandfather. He was even more perplexed when I explained that there was a syndrome named after him, for apparently the family was completely unaware of his fame.

Rezsö Bálint, it turned out, hadn't been his real name either, but rather he had been born to a German Jewish family with the surname Bleyer. When they settled in Budapest, it was thought it might be prudent to change the name to a Hungarian one because of the nationalistic sentiments that were so strong at the time. Bálint is the Magyar (Hungarian) version of Valentine, so the syndrome named after him literally translates as Valentine's syndrome. It was when I told Peter that I was hoping to publish the background of his grandfather's celebrated case that he offered me a photo of him. To my knowledge this is the only one that has ever appeared in print, which happened when eventually we wrote up his story and that of how he came to report his patient.[11]

Bálint's notes were the first description of the symptoms that occur with bilateral parietal damage. More were to follow shortly, but this time from studies of young men who had suffered bullet-wound injuries in the First World War. The velocity of bullets fired from rifles at that time[12] was fast enough to penetrate the skull but, unlike modern firearms, not so fast that they produced potentially lethal cavitation and shock waves in the brain.[13] As a result, soldiers who were unfortunate enough to be shot in the head could survive, and they proved to be an invaluable source of understanding the localisation of brain function.

Most prominent among the accounts of bullet-wound injuries were those of the British neurologist, Gordon Holmes, then attached to the British Expeditionary Force in France. Stationed in the field hospital in Boulogne, Holmes was attending every day to up to three hundred wounded young soldiers brought back from the front-line trenches. Despite the extremely arduous conditions in which he was working, Holmes managed to examine and, most remarkably, carefully record his observations.

What soon became evident was that the design of the Brodie 'Tommy' helmets worn by the British soldiers had a crucial design flaw. Although cheap and easy to manufacture from a single stamping of metal, they sat high on the head, exposing the back of the skull. Holmes's descriptions of young men with focal bullet injuries during the war led him to make some of the most important contributions to our understanding of several brain areas – all of them situated towards the vulnerable back of the brain. They included the primary visual cortex (the first sensory area in the cortex devoted to visual processing), the cerebellum (a key brain region for controlling balance and movement), and the parietal cortex. How Holmes managed to achieve this in the primitive and challenging conditions of a field hospital remains difficult to comprehend, but he had a reputation as a practising neurologist for being both fast and thorough. Indeed, several aspects of the modern neurological examination can be traced back to his techniques.

One of Holmes's important case series from this time were his observations on soldiers with loss of part of their sensory visual field. They literally could not see anything in a sector of their vision, even though their eyes were intact. The reason they were blind in part of their visual field was because a portion of their primary visual cortex – the part of the brain that receives signals from the eyes – had been damaged by a bullet which had passed through it. By mapping the visual field loss and correlating it with the path of the bullet, Holmes was able to establish not only that the primary visual cortex was right at the back of the brain, in the occipital lobe, but also how a map of the visual fields was represented.[14] Objects that appeared in the right visual field activated the primary visual cortex in the left hemisphere, while those in the left visual field stimulated the primary visual cortex in the right hemisphere.

Holmes described another group of soldiers who were suffering from what he termed 'visual disorientation'. These young men had injuries from bullets that had passed through the parietal lobes, sometimes entering one and exiting through the other side of the brain.[15,16] Superficially, they appeared to some people to be blind because they would stumble into objects, bump into people and not know where things were when asked to pick them up. But, as Holmes discovered when he examined them carefully, these soldiers weren't blind. They could see, though typically, when asked to touch an object they might move their arm in the wrong direction and continue to grope for it 'more or less like a man searching for a small object in the dark'. Holmes demonstrated that his patients simply didn't know where objects were located in space, and some of them also suffered severe inattention. His descriptions are very similar to the patient Bálint had described with strokes affecting the left and right parietal cortex. Hence, nowadays, neurologists refer to the constellation of visual deficits that these pioneers reported as the Bálint-Holmes syndrome.

Others, too, were recording observations of curious symptoms as a result of parietal injury. German soldiers who were shot in the front lines of the First World War were examined by neurologists such as Poppelreuter and Kleist.[17] The latter introduced the term 'constructional apraxia' to denote the impairments in copying three-dimensional constructions or drawings that parietal injuries led to. Later, towards the second half of the twentieth century, it slowly became apparent to many neurologists that visuospatial syndromes such as neglect and constructional apraxia were far more common and more severe after right-hemisphere than left-hemisphere damage.[2,18,19] Exactly why this should be the case remains to be established, but one explanation is that because the left hemisphere in human beings has become so specialised for speech and language, many other functions – particularly spatial ones – became localised in the right hemisphere.

~ o ~

Winston came back to see me in the clinic a month later. He was dressed altogether differently: no suit on this occasion but a slightly threadbare white shirt and some bottle-green tracksuit bottoms, which seemed very incongruously put together for a smart dresser like him. I could see that he was still suffering from left-sided neglect. He had shaved the right side of his face neatly, but there was clearly untended grey stubble on the left, and on his left sleeve there were several yellowing stains, perhaps the vestiges of culinary endeavours over the last week. He looked despondent but tried to be upbeat.

'What's up with you, doc?'

'I'm well, thanks, but how about you?'

'Not doin' too badly,' he said, nodding his head in a very unconvincing style.

'No? You're not having any difficulty?'

'Well, we all got difficulties, man!' He laughed. 'Who doesn't?'

'But after the stroke, I mean.'

'Ah well, ya sure I did have a stroke, doc?' He was looking me straight in the eye, which was definitely an improvement since I'd last seen him. Previously, he would always be gazing off to his right.

'Yes, I'm sure. Any reason to doubt it?'

'Well, my friends doubt it.'

'How do you mean, Winston?'

'Well, they say they haven't seen people with a stroke like me. Me walks fine. Me talks fine. But I'm still bumpin' into people. My friends, they think there's sumtin' else going on. They're, how d'ya say, reluctant to go out with me. They don't say as much, but I can sense it.'

'I see.' I nodded. 'And what do you think about the stroke?'

'I trust ya, doc, but I don't seem to be gettin' any better.' There was still a sceptical tone in his voice.

'It's only been a couple of months since it happened, so it's still early days.'

'Really? Isn't there any medication or rehab for it?'

'I'm afraid there isn't any licensed treatment for this neglect syndrome that you have.' And at that time there weren't even any clinical trials that we could offer to recruit Winston into. Several years later, we were able to show that guanfacine, a drug that boosts the actions of the neurotransmitter noradrenalin in the brain, can significantly improve attention in some people

168

with leftward neglect after their right hemisphere stroke.[20] Drugs that simulate the effects of dopamine in the brain can also boost attention in some patients.[21, 22]

'Huh, so all I do is wait?'

'A lot of people do get better over time.'

He nodded his head again. I changed the subject.

We started to talk about Winston's background. He was one of the 'Windrush Generation', so called because of the *Empire Windrush*, an ex-German cruise ship, which in 1948 travelled from Jamaica and other Caribbean islands, bringing some of the first migrants who were seeking work in the UK. The British government had actively encouraged people from across its former Empire to move to the country, because there was a shortage of labour in post-war Britain. Winston had crossed the Atlantic in the 1950s. However, what soon become apparent to him and his compatriots was English people's hostility to Black people. They were not welcomed with open arms.

'Whenever we went for a job, we'd be told it had just been filled. When we tried to find a place to live, people might just straight out say that they didn't take "wogs" (a racial slur, which many take to be a shortening of golliwog). Sometimes there were signs outside: "No Blacks, No Irish". Now, what chance did I stand when I'm both of those, doc!' He laughed. 'Ya know, even when we did find a room to rent, the landlord would complain about the smell of our cookin': "We don' want no darkie food," they'd say.'

Like many others from the Caribbean, Winston had settled in Notting Hill. Now a fashionable part of west London, at that time it was the very opposite. Exploitative landlords had carved out some of the old houses there into multi-occupancy flats and, because they were prepared to rent to Black people, many people from the Caribbean in this wave of migration ended up living there. It was a part of London where illegal drinking and gambling, prostitution and drug dealing were rife. It was also the focus of festering tensions between the new migrant community and the locals.

'I was there when the big riot happened in 'fifty-eight,' Winston recalled. 'It was crazy, mad stuff with Teddy Boys attackin' people in the street with iron bars and knives for no good reason. We had to fight back.'

He was talking about a group of young, suburban working-class men who'd taken to wearing clothes based on Edwardian-style fashion (hence 'Teddy'). Although one of the interests of these Teds was rock and roll music, another seemed to be living a violent lifestyle which notoriously came into public view in the Notting Hill riots of late summer 1958 that Winston was referring to. These were triggered by an assault by a gang of white youths on a white Swedish woman who was married to a Jamaican. The suspicion was that the subsequent eruption of violence involved Oswald Mosley's far-right Union Movement and the White Defence League.

Mosley, who had been interned for much of the Second World War for his fascist views, stood as a candidate for Member of Parliament in the 1959 general election in the Kensington North constituency, campaigning stridently on a platform of the forced repatriation of immigrants from the Caribbean. His manifesto included the prohibition of mixed marriages. In the end, he was unsuccessful, securing only eight per cent of the vote. In response to the racially motivated riots, an indoor Caribbean carnival, televised by the BBC, was arranged by Claudia Jones, a journalist and activist from Trinidad. This was the precursor to London's most famous street event: the Notting Hill carnival.

Winston was looking down at the floor.

'It must have been a tough time for you.'

He looked up. ''Twas worse than tough, doc. We felt lucky that we hadn't been lynched. Many's the time I thought about

170

going back to Jamaica, but we hung on and tings improved, I gotta say that.'

Winston's descriptions of Notting Hill brought back some images for me as a child growing up in west London in the 1960s. My first encounter with a lot of Black people from the Caribbean was in Shepherd's Bush market. Holding my mother's hand, I would walk tentatively with her across the Goldhawk Road towards this noisy, bustling arena, which my parents would insist on calling a bazaar. As we entered, we would be greeted by a cacophony of cries from the market sellers – 'Cum'n'gettum. Luvvlyapples. Two'n'six. Cum'n'gettum' – all unfathomable to the ears of an immigrant child. Most of the people with stalls were white Londoners. But to one side of the market, there were the Black Caribbean traders with okra, sweet potato, green bananas and vivid varieties of fish that I'd never seen before: red snapper, whiting and mullet. Most of all, I still recall the exotic smell of curried goat being served by a guy sporting a trilby, shouting out for customers in an accent that was even more difficult for me to decode. I didn't know what he was saying, but I knew I wanted to taste his food.

'Tell me about your friends,' I prompted Winston.

'Well, I gotta a few really good ones. I've known 'em since those days. We stick together. Well, we did, but only Kelvin comes around nowadays.'

'You married?'

'No, man, but I gotta son. He lives in Smet'wick, up in Birmingham. Don't see much of him.'

'So, your friends don't think you've had a stroke?'

'I'm not makin' it up. Ask Kelvin. He's out there in the waitin' room.'

'Oh, I hadn't realised he's here with you. Do you mind if I talk to him alone for a few minutes?'

Winston agreed and I guided him towards a seat in the

waiting area where I also spotted his friend, who rose to shake my hand.

'Good to see ya, doc.'

'And you, Kelvin. Thanks for coming with Mr O'Hara.'

'I couldn't do anytin' but. He needs help, doc. He wouldn't 'ave made it here on his own.'

I took him into the consulting room to speak to him alone. Everything about Kelvin's manner was gentle, restrained and understated. Winston was obviously the extrovert of this partnership; Kelvin the quieter foil.

'How's he doing back home in Notting Hill?' I asked.

'He's far from right, doc. Winston, ya know, is the life and soul of the party. He's the man we go to see for fun. He's what makes us laugh but now, tell ya the truth, we get a likkle embarrassed goin' out with him. He's never safe on his own. The moment he starts to walk, he's always bumpin' into people or knocking tings over. When he's eatin', he'll not notice that he's dropping food. Ya can see the stains on his shirt. His place is a tip 'cause he's not able to look after it, like. That's not him. Him a very proud man. Some of Winston's friends have got a bit suspicious, like. We never seen anytin' like that in people who had a stroke. Ya not hidin' sumtin' from him?'

'How do you mean?'

'Well, we got to talkin', his friends see, and we were tinkin' that maybe Winston has sumtin' terrible that you're not tellin' him.' There was a pause. I looked quizzically at Kelvin.

'What do you mean?' I asked.

'A long time ago, one of our neighbours, he was diagnosed with syphilis that had gotten to his brain, even when he was an old man. He went a bit doolally. Sumtin' they called general paralysis or some such.'

I was a bit taken aback. Kelvin was referring to what in the nineteenth century had been called 'general paresis of the insane', a term that still remains in medical practice today. GPI, as it is now abbreviated to, is a late manifestation of untreated syphilis. Characterised by progressive restlessness, decline in memory, personality change and often delusions of grandeur, patients with the condition are observed to slowly become unbalanced, weak and tremulous before being unable to walk and, eventually, dying bedridden.

At one stage up to twenty per cent of men in British asylums in the late nineteenth century were given a diagnosis of GPI.[23] Some have speculated that Robert Louis Stevenson's character Mr Hyde was based on someone with GPI. Although it wasn't established at the time that syphilis was the cause, the fact that most patients with the condition were men did not pass un-noticed. Finally, in the early twentieth century, the discovery of the bacteria that caused syphilis in the post-mortem brains of patients who had suffered GPI left little doubt as to the cause. The discovery of penicillin had a major impact on treating the condition, provided it was not diagnosed too late.

I knew that GPI comes on slowly over years, not suddenly as in Winston's case, so there was no reason to believe he had this condition. However, there was one complication: sometimes syphilis infection in its intermediate stage can cause an inflammation of the blood vessels, including those in the brain, which in turn can lead to recurrent strokes. But Winston had not suffered multiple strokes, and our trainee Nicky, who was nothing if not extremely thorough, had asked for a syphilis test to be performed on the blood sample that Winston had given. Thankfully, it had been negative.

'Oh, I'm sorry to hear that, but no, I haven't got any reason to think Mr O'Hara has anything like that.'

'Ya sure now, doc? Cos that's partly the reason that some of our friends aren't seein' Winston. They're worried they might catch this disease. I ain't told him nuttin', though.'

I looked at Kelvin in some surprise. This was totally unexpected for me.

'Let me bring him back into the room,' I said, 'and perhaps I can show you both the stroke on his scan if he is happy for me to do that.'

'Okay, let's do that.'

And so, I brought Winston back into the room and he agreed that it would be alright for Kelvin to see his brain scan with him.

'At least he'll believe I have a brain!' he said and chuckled loudly.

I logged into the radiology program and found Winston's CT scan.

'Here, you see the darkness here, that is the stroke,' I said pointing to the screen. 'You see the other side of the brain, the left side, is normal. There is no darkness there.'

'Darkness, ya say,' Winston mused as he looked at the scan. 'And that's my brain?'

'Yes, all the scan shows is a stroke. The blood tests don't show you have any kind of infection either. If we control your blood pressure and you take aspirin, we can try to reduce the chances of you having any more strokes, Mr O'Hara.'

'And this problem I have, it's called neglect, that right?'

'Yes, and many people do get slowly better over six months.'

'And that's all there is, doc?' Kelvin enquired too, looking for reassurance. 'No infection, or nuttin' like that?'

'Absolutely no infection,' I emphasised firmly.

'Funny ting, this neglect. Ain't it a queer ting, Kelvin? Anyone doing research on it?' Winston asked.

'As a matter of fact, we are. We're trying to understand what happens in people like you: why you're not paying attention towards one side of space and missing things on that side.'

'Really? I'm happy to help if you need me, doc,' Winston offered.

'Well, ya don't wanna be a guinea pig for research now though,' Kelvin said, swiftly intervening.

'You wouldn't be a guinea pig, I promise. We'd do some special tests of your vision, attention and memory, and perhaps ask you to have another type of brain scan.'

'I'd be up for that, so long as ya look after me, doc.'

~ o ~

When Winston left, I couldn't help but think about his predicament. Here was a man who had been at the centre of his Notting Hill group of friends, but now he was facing being ostracised because of his inattention and the suspicion that this wasn't due to a stroke but rather syphilis. I decided that we needed to get him involved in our research sooner rather than later, so that his friends would know that he was taking part in work that might help patients who had suffered a stroke. The next day I called Kelvin to see if we might arrange a taxi to pick Winston up and bring them both to the research centre at the Institute of Cognitive Neuroscience at Queen Square.

'So, this is where the guinea pigs come, right, doc?' Winston smiled.

'No, this is where patients with stroke who have neglect come for our research studies.'

'So, what ya goin' to do me, doc? None of that electrical stimulation I hope!' He chuckled. 'I have had too much stimulation in my time.'

'Let's take you to the brain scanner,' I said.

'Now ya talkin'. We gonna get some fresh photos of my brain, Kelvin. Bet you're jealous!'

'And this research is for stroke patients, right, doc?' Kelvin asked.

'Yes, just for stroke patients,' I said emphatically. 'Nobody else.'

In the MRI scanner, Winston lay looking at a central cross

while we flashed visual images either to the left or right of it. We got him to press a hand-held button whenever he saw something. The scans of his brain revealed Winston's right parietal stroke, as previously. They also showed that the region of his primary visual cortex was intact in the right cerebral hemisphere, just as it was in the left. This explained why his visual fields were intact.

Later, when we analysed the activity in his brain, we found that whenever an image was flashed in his right visual field it would activate the primary visual cortex in the left primary visual cortex. Similarly, when we presented an object in his left visual field, there was a response in his right primary visual cortex. We could tell that this occurred even on the occasions when Winston failed to be consciously aware of the image that we flashed in his left visual field.

This type of finding was scientifically extremely important. It showed us that even though the visual cortex in the right hemisphere was activated by an object in the left visual field, this was not sufficient for conscious awareness of it.[24] Although his eyes saw an object to his left, and even when his primary visual cortex 'saw' it too, Winston's left-sided neglect due to his right parietal stroke left him unable to perceive it consciously. He wasn't able to see this object. Thus, activation of the primary visual cortex is not sufficient for human beings to perceive with awareness.

One possible conclusion from this sort of finding is that we need to be attending to something to be consciously aware of it. John Marshall and Peter Halligan, two Oxford neuropsychologists, provided further evidence for this. They asked a patient who, like Winston, suffered with left-sided visual neglect to look at drawings of two houses which were identical – except one of them had flames and smoke coming out of its left side. When asked if she noticed any difference between the two buildings, the patient reported that they were identical. She failed to notice that one of the houses was on fire because of her left-sided inattention. However, when she was requested to indicate which of the houses she would prefer to live in, she consistently chose the one which was not on fire.[25] Even though she was not consciously aware of the fire on the left side of one of the houses – and she did not report seeing it – this information had entered her brain and was biasing her choices subconsciously.

Before Winston left the research centre with Kelvin, he asked to say goodbye to me.

'That flashing stuff in the scanner, are you really tellin' me that helps you understand what's goin' on in my brain, doc?' he asked, obviously bemused.

'Yes, we hope so. Thanks for coming all this way to take part in this research, Mr O'Hara. Without people like you being involved, we wouldn't be any the wiser.'

'Please call me Winston,' he insisted. 'And, doc, ya sure about there bein' no infection that's caused this?' he asked once more.

'Is there something in particular you're worried about, Winston?'

'Well, ya know, one of our neighbours went a bit funny and it turned out he had syphilis. And people are puttin' two and two together and making six out of it.'

'I see,' I said, quietly recognising it as the story Kelvin had told me too. 'We actually tested you for syphilis when you first came to hospital and it was negative. The scan doesn't show anything else suspicious on that front, so there's no need for you to worry about that.'

'Ya already tested me for syphilis.' Winston beamed. 'Ya crafty people! Well, ya made my day, doc.'

I waited a minute.

'Winston, I couldn't say this to your friend, but would it help if I told him about the test result? Would I have your permission?

177

If Kelvin heard it from me, he might be able to reassure your friends.'

'Ya'r brighter than you look, doc! Of course, please do tell him. It would really help.'

So, I brought Kelvin over and explained. He was also very pleased.

'Well, that's a relief, doc. If I can say that to our friends, that might make a likkle difference.'

And it did make a difference. Winston's friends were no longer anxious about catching anything from him, although at times his inattention still rankled. They interpreted it as confusion and its manifestations in public meant that, to begin with at least, they were still unkeen to go out with him anywhere. Over the next six months, however, Winston's neglect slowly improved, as it often does in patients with stroke. Although it never fully resolved, it became possible for him to reconnect better with his circle of friends. He was able to go out with them, sometimes to the pub, where he no longer bumped into people or knocked over glasses, sometimes to a club, often to Portobello market or the local park. He even managed to enjoy again the vibrant street life of the Notting Hill carnival, an annual event that he had been to for over thirty years.

We saw Winston a lot over subsequent years. He took part in a whole host of investigations on the functions of the parietal lobe in short-term memory, sustaining attention over time and even directing eye and hand movements to objects at different spatial locations. Of all the patients I have seen, Winston contributed more than anyone to our research. And each time he came to see us, he would bring a different story from Notting Hill, the place he loved to belong to.

6

The woman who didn't care

Queen Square was looking eerily resplendent. Unusually, the winter sun had broken through the cloud cover in places, framing the barren trees in a soft pink light and imbuing the scene with a wondrous stillness. The overnight frost still held tight to the branches, etching them in a silvery ink. The gardens were serenely quiet. Moments like these in the heart of London are to be cherished, and as a treat I decided to sit down on a bench to absorb the atmosphere. A few feet away a splendid red-throated robin, curious to inspect the newcomer, hopped closer to make my acquaintance, but was soon evidently satisfied that I offered little of interest. I was sure that he was about to fly off when we were both startled by shouts from the roadside. The peace was shattered.

'I am not going to take that from you!' I heard a high-pitched voice scream.

'Madam, I'm very sorry, but you asked for a cappuccino and that's what I made.'

'Do you know that you're an absurd specimen of a man? If we can call you that! This tastes like drain water. Not any old drain water but one with foul, frothy scum on top. And you have the nerve to call it a cappuccino. It deserves to be in the bloody sewers.'

The voice echoed harshly through the stillness. Curious to see what was going on, I stood up to look through the garden gate. In front of a little van which was selling coffee stood a petite woman, no taller than five foot, perhaps in her late fifties. She was dressed in a suede pink cowgirl suit, with sparkling silver tassels. On her head was perched an oversized white Stetson, which was in danger of engulfing her entire face. On her feet she wore white high-heeled crocodile leather boots, embroidered with a lurid depiction of an alabaster serpent coiling upwards. It was an extraordinary sight.

'I'm really very sorry, madam. I'll make you another one.'

'You must be joking! I don't want any more of this filth.' With a flourish, she poured the milky contents of her paper cup straight into the gutter and marched off indignantly. Behind her, an apologetic man mumbled something to the barista, shrugged his shoulders and quickly followed his cowgirl companion. Thankfully, the shouting had ceased. Quietness had returned to the Square, but my moment of enjoyment had been spoiled and I had a clinic to get to.

The list that morning was busy. It started with me seeing a patient who had suffered from physical symptoms due to her multiple sclerosis for more than three decades, but now she was also evidently experiencing a decline in her memory. The question posed by her GP was whether her cognitive impairment could all be attributed to her multiple sclerosis (certainly a possibility), or whether she had unfortunately developed a second condition (such as Alzheimer's disease). We had to arrange neuropsychological testing and an MRI brain scan as the first set of tests to distinguish between these possibilities. Next, I had the difficult job of giving a man in his sixties the news that he probably had an unusual neurodegenerative condition called progressive supranuclear palsy. Neither he nor his family were expecting this. I therefore had to go beyond the time normally allocated for a consultation in order to explain more about the condition.

As I was finishing the letter for this consultation, the clinic nurse entered my room, clearly flustered.

'Are you ready to see your next patient?' she asked breathlessly.

'Sure, but just give me five minutes to finish this letter.'

'Would you mind leaving that until later? The next one on your list is creating a bit of a stir in the waiting room. They're extremely unhappy that you're running late.'

I looked at the clock above the door. 'But I'm only ten minutes over.'

'I know but they're making our lives hell out there.'

I nodded. 'OK, bring them in.'

I put the Dictaphone down. Patients had become so much more demanding over the years, I reflected. You might experience more than a ten-minute delay to see a hairdresser, yet it was seemingly unacceptable when your appointment was to see a specialist doctor. Why such a fuss?

When they came in, I realised the answer straightaway. There stood the cowgirl and her companion from the squabble I had witnessed that morning. I took a deep breath in.

'About bloody time! I'm Sue Ryland,' she exploded.

'Hello, I'm . . .' I was starting to introduce myself, but she didn't give me a chance.

'I don't care if you are sorry. You're late, and if there's one thing I cannot stand it's people who are late,' she interrupted. 'Don't look at me like that. You doctors deserve to be told, just like everyone else. It's impolite to keep people waiting.'

She was genuinely livid. Her white Stetson now in her hand, I could see her rounded face, framed by unnaturally jet-black hair and expressing a deep rage which I did not really want to confront. I stayed quiet, waiting to see what would happen next, but it was he who unexpectedly spoke.

'Hello, doctor. I'm very sorry. Please excuse Sue. She wouldn't normally behave in this way. That's part of the reason we're here to see you. I'm Alan, her husband.'

He was a short man too, but with a freckly face and thinning ginger hair. His wiry frame contorted stiffly with embarrassment as he apologised for his outspoken wife.

'I see. Good to meet you both. Please do sit down,' I responded, keeping my words to a minimum in an attempt not to inflame the situation further. I got the impression that saying anything might provoke Sue.

She was now looking at Alan with exasperation, her focus turning away from me.

'Charming! How dare you even think of apologising for me, Alan! He's the one who should be saying sorry,' she said, stabbing her finger in my direction. 'He's the one who should apologise, Alan. Not me.'

I waited again.

'I really am sorry to have kept you waiting,' I said. 'I had to give some bad news to someone. I hope you'll understand. That can take longer than our normal appointments. But I'm seeing you now and will give you the time you need.'

'I should hope so,' said Sue, suddenly bringing her feet off the floor to sit ankles crossed, with her crocodile boots perched on the edge of my desk.

'I'm so sorry, doctor. Sue, please take your feet off the desk,' implored Alan, clearly even more embarrassed.

'Why? There's nothing wrong with it and this doctor wants me to relax, don't you?' She smiled at me wickedly, fully sensing my awkwardness.

'That's fine with me,' I said slowly. 'If you find it easier to speak with your feet up on the desk, I don't mind. By all means, Mrs Ryland.'

'See, he's alright with it.' She nodded to Alan, who had a look of resignation on his face. 'You never know, he might be a bit of alright too,' she commented as an afterthought, laughing out loud. Alan squirmed.

Now Sue was a little calmer, I tried to take a history from her.

'Well, it's true what Alan says: I am different. I have changed, but it's all for the best, you know. I used to be one of those timid little housewives, but I've had enough of all that. The kids have grown up. Alan's close to retirement. I can just be myself now. I don't care what people think,' she explained.

'I see, so how have you changed?' I asked gently.

'Well, I do what I want. I don't have to pretend.' She drummed a short tattoo with her fingers on the arm of her chair, in the way someone might if they were impatient.

'Give me an example.'

'Where do I start? I've taken up line dancing. I never would have done that. Do you like my outfit, by the way?'

'It's certainly very arresting, Mrs Ryland,' I commented.

'Thank you. I thought you doctors might appreciate being cheered up a bit. You're such stiff, gloomy people. Alan had suggested I wear something more boring, but it's nice to be valued, I find. Did you hear what he said, Alan? He likes it.' Alan kept quiet, his eyes looking up to the ceiling as she continued.

'Then I've built a home cinema. Always wanted one, but everyone kept saying there was no space, so I've got it built in the garden. It's amazing, isn't it, Alan?'

'Yes, we now have a home cinema in a shed in the garden, doctor. Sue insisted and we couldn't refuse.'

'It's been magic. I watch some great movies in there. We've got surround sound and everything. Top notch. You can come if you behave!' She drummed her fingers again.

'Thank you. Would people outside the family notice that you're different, Sue?'

'I should bloody well hope so. I'm not being all meek and demure anymore. I just speak my mind, but there's nothing wrong with that, doctor. There are too many things left unsaid in this world.'

And so it went on. Sue was insistent that she was just being herself after years of conforming, of playing the role expected of a mother and wife.

I asked her to step out the room for a moment, so I might get the perspective of her husband and she agreed, but not without a little reluctance.

'Don't you go telling any of your porky-pies, Alan. I don't want this flirty doctor to get the wrong idea of me before we have a chance to hit it off,' she said with a chuckle as she left.

'Is what she says correct?' I asked Alan.

'Oh, it's true alright, doctor. Too true.'

'When did things change?'

'I first noticed about two years ago. We were walking down the high street and she saw this couple. He would have been in his seventies. She was in her twenties. He put his arm around the shoulder of the young woman to give her a hug, and before I knew it Sue was shouting, "Filthy bugger, you should be ashamed of yourself, taking advantage of a young girl like that!" They didn't know what to do. Both of them were very upset. He turned out to be her grandfather taking her out for a birthday lunch. Sue would never have shouted out in public like that in the past, but now she's commenting on everyone. The other day, she saw a woman across the road and called out, "You need to lose weight, love. That dress is doing you no favours." That isn't my Sue, doctor.'

'So, has her behaviour got progressively worse over time?'

'Oh definitely, much worse.'

'What was she like before then?' I asked.

'Well, you wouldn't believe it from what you've just seen, but she used to be a quiet, understanding person who everyone loved to talk to. The neighbours go indoors now if they see her coming down the road, because she can sometimes get very agitated, you know – verbally aggressive. In the past, they'd chat to her for ages because she was always so helpful. To tell you the truth, I cringe being out with her. For example, this morning she gave this poor coffee seller such a hard time, he didn't know what hit him.'

'I know. I saw it happen,' I remarked.

'Did you, really? Well then, you know what she's like. That's a prime example. But the worst one was two weeks ago. We were on the bus and there was nowhere to sit because it was so crowded. What does she go and do? She squeezes a young man's bottom, saying "You'll go far, my lad, with an arse like that." He was shocked; he didn't know what to do but made sure he got off at the next stop. I wished I could have too. People were looking at me, disgusted, as if I was responsible. And then all Sue said to them as this young man got off the bus was, "What! Don't look at me like that – he does have beautiful buttocks!" I could have died.'

'And she never would have done that in the past?'

'Oh, my goodness, no. She was prudish, if anything.'

'So why did she say she'd done that on the bus?'

'Her response was that there isn't enough plain speaking in the world. He was a very good-looking bloke, and she thought she'd show her appreciation. I mean, I ask you!'

'And how is she with the family?'

'Well, she used to be a really warm person, very affectionate, always hugging me and the children. Now she seems, well, very disengaged, quite cold. She doesn't seem interested, never seems to show any feelings or emotions when there's good or bad news. For example, our daughter had her first baby – he's our first grandchild – two months ago. Sue didn't show a flicker of interest. I had to insist that we go to visit our daughter after the baby was born. "Must we? There's a good film on the television tonight," she says. Can you believe that? This is our first grand-child. Sue doesn't seem to care if she upsets people either. The other day, our daughter was crying on the telephone because she wasn't coping with the baby and Sue just told her to get a grip. "We've all had to deal with it," she said. "I can't be doing with any moaning or dramas now. You're not the first woman to have a baby." And she just put the phone down. She no longer seems to understand people, even her own family. She lacks . . . what's the word . . .?'

'Empathy?' I offered.

'That's it. She lacks empathy.'

'How else has she changed?'

'How hasn't she changed!' blurted Alan in exasperation. 'She told you about the home cinema. None of us wanted it except for her. We don't have any space, but she insisted, even when the children were against it. She'd never have done that in the past. Now we have a huge shed that takes up most of our garden where she disappears to watch movies.'

'Is she a big film buff?'

'No.' Alan shook his head in bemusement. 'That's what caught us by surprise, but she says that she never had the courage to say what she wanted before. It's all beyond me.'

'Is she doing anything else that is out of character?'

'Huh,' he sighed. 'Well, in the last two years, she has bought eight different bikes.'

'Why so many?'

'That's just it. *Why so many?* She insists she needs a different bike for different situations. You know a mountain bike, a racing bike, a folding bike and, would you believe it, a tandem was her latest purchase. And she hardly rides any of these bikes.'

'And how's her memory?'

'Sharp as a button. It's better than mine. She doesn't forget, but she is now so much more disorganised. Sometimes she'll start to do something in the house, like reorganising the kitchen cupboards, but then she'll get distracted and go off to do a bit of gardening. But before she's finished that, she might go off to the computer to order something we don't need, leaving the kitchen and garden in a mess. She'd never do that before.'

'And is there anything else that you think is important that we haven't talked about?'

Alan suddenly looked very embarrassed. I gave him time.

'There is one thing,' he said awkwardly.

I waited.

'A few weeks ago, we were in the supermarket. I'd gone off to get some milk or something and, when I was walking up to Sue from behind, I saw that she was putting chocolate bars into her handbag – four or five of them. She was going to shoplift them, and I wouldn't have known anything about it. She didn't even seem to feel guilty when I fished them out and put them into our shopping trolley. She would have been mortified if she'd been caught doing that kind of thing in the past.'

All these changes in Sue's behaviour were, understandably, extremely concerning to Alan. She would blurt out what she thought, she acted impulsively, sometimes she could be verbally aggressive, she lacked empathy, had poor judgement, seemed to be chaotic in how she did simple everyday things at home, and had even been caught in the act of shoplifting but showed no remorse. The only other thing he could think of was that she had developed a habit of repeatedly drumming on a table with her fingers.

'You mean like she was doing in this room?'

'Yes,' he agreed with a nod, 'but that was mild. At home she can keep repeating this over and over again. It's annoying.'

'Is there any family history of anything similar?'

'Not as far as I know. Her mum and dad are in their eighties

187

and they're doing well. She has two older brothers and a younger sister. None of them have anything like this.'

'And how is Sue's relationship with them and her friends?'

'Awful, just awful. She's had rows with her parents and one of her brothers over silly stuff, like politics of all things. She's become so rigid in her views. She has no tolerance. She can't stand anyone saying a kind word about immigrants or gay people, or anyone she thinks is different really. She never was like that. Her brother won't speak to her after she was rude to him about his wife. As for friends, we hardly see anyone nowadays. She's even managed to upset a few of my old mates, so they won't come around to our house. I have to meet them elsewhere. Frankly, she's become a nightmare.'

I brought Sue back into the room. The neurological examination I performed did not reveal any abnormality, but she was not too happy about undergoing the process.

'As if tapping a few reflexes is ever going to tell you anything!' she observed.

The cognitive testing screen I performed didn't show very much either, except that her verbal fluency (the ability to think of words beginning with a letter of the alphabet in a minute) was reduced. However, just as my conversation with Alan had indicated, her episodic and semantic memory were both very good, as was her attention and visuospatial function.

'I think we'll have to move to a scan of the brain and some deeper tests of your cognitive functions with our neuropsychologist,' I explained.

'Well, that is a surprise,' Sue responded sarcastically. 'And what if those are normal, too? You'll have to say there isn't anything wrong with me. I'm a liberated woman, doctor. That's going to be your diagnosis.'

~ o ~

It was a relief when Sue finally left. She had been quite exhausting. It had been challenging to keep calm in the face of some of her provocative remarks. I had felt myself getting very irritated, occasionally even a little angry, throughout the consultation. That was telling in itself. Few people have the power to cause such an emotional response in most of us, but Sue could do it, although apparently she had not behaved like this two years ago.

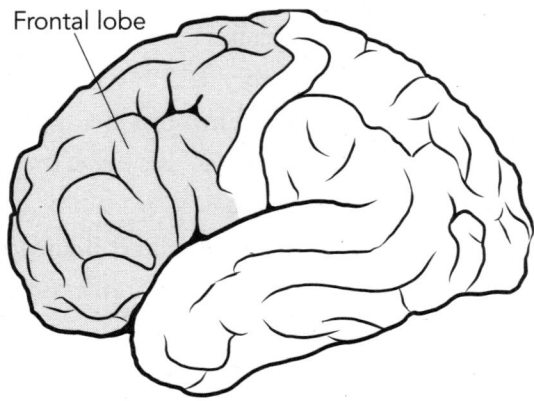

Fig. 13 | Frontal lobe. The frontal lobe is the largest lobe of the human brain.

Her behaviour was highly characteristic of patients with what has been termed the 'frontal lobe syndrome'. People with injury to their frontal lobes (**Fig. 13**) can become very different in their behaviour. This was first appreciated in the famous case of Phineas Gage,[1] a foreman who was working with his men in September 1848 on the construction of the Rutland & Burlington railroad in Vermont. Gage was preparing to blast a hole through a rocky outcrop by compacting gunpowder that he had poured into a hole in the rock with a tamping rod. Unfortunately, the powder exploded, and the force sent the metal rod flying straight through his skull. He was lucky to survive, but his subsequent progress was extremely stormy. He required surgery for a brain abscess and was fortunate enough to be attended by Dr John Harlow, a New England doctor, who had the experience to perform this.

Although Gage survived, his personality was profoundly altered. From being a responsible, upright man in his local community, he became an irritable, impatient person, prone to swearing and unreliable at work. Sometimes obstinate and rigid in his views, and at other times highly capricious or vacillating in his judgements, he could seem callous and aggressive. His friends no longer considered him to be the same person they had known. Unable to hold down his previous job, Gage moved to Chile but when his health began to decline, he returned to the States, eventually dying from a prolonged seizure in 1860 in San Francisco. When Harlow found out about his unusual patient's demise, he persuaded Gage's family to have the body exhumed so he could examine the skull. This showed the entrance and exit hole of the tamping iron as it had passed through the frontal lobe of the unfortunate man's brain. It is still viewable in the collection at Harvard Medical School today.

The publication of Gage's case led to his importance as the first, well-documented individual with behavioural changes after frontal lobe injury. Since that time there have been many descriptions of the effects of frontal dysfunction on people's personalities and behaviour. Patient EVR, for example, reported by Eslinger and Damasio in 1985, was a high-achieving professional. He was made chief accountant of his firm when he was only twenty-nine.[2] At the age of thirty-five, he was diagnosed as having a very large, slowly growing benign brain tumour called a meningioma. This was resected surgically, but although the operation was successful, EVR appeared very different afterwards.

Impulsively, he invested all his savings into a risky venture and soon became bankrupt. Subsequently, he drifted through several different jobs, including working as a warehouse labourer and building manager, but was fired from all of these. Employers complained about his poor timekeeping and disorganisation. Apparently, he might take two hours to get ready for work.

190

Some days were spent entirely on shaving and washing his hair. If he was planning to have dinner out at a restaurant with friends, it took him hours to decide where to sit or to select something from the menu. Frustrated and unable to cope, his wife left with their children, divorcing him after seventeen years of marriage. He moved in with his parents. Within a month of his separation, he remarried against the advice of relatives, only to divorce again two years later. Despite the clear alteration in his behaviour, when tested on the standard cognitive tests that were available at that time, EVR performed very well. This led many researchers to conclude that those tests did not directly capture some of the functions of the frontal lobes that are evident in the day-to-day lives of patients like him.

Some individuals with frontal lobe dysfunction might be very impulsive or disinhibited when they interact with other people. Sometimes they are rude or overfamiliar, as appeared to be the case in Sue. Others can become very withdrawn and lack initiative do anything. Some become highly rigid and inflexible in their thinking. Sue apparently showed some of this in her behaviour. Some patients have difficulty following instructions, learning new rules or solving relatively simple problems, such as how to use a device like a television remote control. Acquaintances might comment that they no longer seem able to coordinate activities or multitask. Or they might experience difficulties planning ahead or deciding on which action to prioritise if they have many things to do. Some may become highly distractible, unable to stay 'on task'. Some patients become socially or emotionally inappropriate, for example, no longer empathising with others, or being able to see things from another person's point of view, as also appeared to be the case in Sue. The precise constellation of symptoms can differ widely between patients.

The myriad of behavioural changes that can occur in the frontal lobe syndrome has been extremely puzzling to neuroscientists.

Hans-Lukas Teuber, Professor of Psychology at the Massachusetts Institute of Technology (MIT), famously referred to this perplexing collection of symptoms in 1964 as 'the riddle of the frontal lobe'.[3] The frontal lobes and the functions attributed to them still remain something of a mystery. If you ask any neurologist, neuroscientist or psychologist what the frontal lobes do, you will receive a different answer from each, but somewhere at the heart of all their responses will probably lie the concept of 'control'.[4] Since the very early descriptions of the consequences of frontal lobe injury, control has often dominated discussions on the topic, so much so that some researchers have proposed that one way to think of what the frontal lobe does is to imagine that it acts almost analogously to the chief executive of a company.[5]

According to this view, it makes decisions, formulates goals, plans ahead, initiates actions or stops ongoing ones if it realises that these are ineffective. If necessary, it might switch to a different strategy to solve a problem. It keeps us focused on the task at hand, coordinates our actions and multitasks over time. It considers how other people view a situation and, if appropriate, empathises with them. It grasps social context and acts accordingly. It may elect to persuade or cajole others when necessary. From this perspective, the frontal lobe plays a critical role in what has been termed 'executive control' – sometimes also referred to as 'cognitive control' – supervising or coordinating the activities of the rest of the brain in pursuit of the overarching goals it wants to achieve.[6]

David Badre, a cognitive neuroscientist at Brown University, argues that in order to achieve a particular aim, the brain needs to have a mechanism to bridge the gap between knowing something about the world (knowledge), and achieving the goal it aspires to achieve (through performing the appropriate actions for a particular context).[4] He observes that some patients with frontal dysfunction fail to achieve their goal even when they can say what they want to achieve. Connecting – or crossing – the bridge between knowledge and executing the relevant actions, Badre proposes, is what frontal control systems allow us to do. In a flexible manner, they allow us to plan and select, to create a sequence of actions and also to monitor the consequences of our behaviours for ourselves, as well as their impacts on other people.

In fact, Badre goes further to argue that the frontal lobe is crucial to successful anticipation of the consequences of our actions. If we foresee that a particular strategy is likely to be unsuccessful in a particular context, we might not execute it. For example, making a joke at someone's expense in a group might be perfectly acceptable when out socially with colleagues, but it may not be appropriate within the same group when discussing an important project in a meeting at work. Different parts of the frontal lobe may play selective roles in the hierarchy of control – from creating a goal, to considering the particular context in which it has to be performed, through to executing the actions required to achieve that goal in that situation.[7,8]

The impulsive, disorganised and chaotic behaviour of individuals with frontal lobe dysfunction has been referred to as the *dysexecutive syndrome* because of the seeming lack of executive control over behaviour. (It's as if the chief executive officer or controller of a large organisation is no longer making good decisions over what the company – in this case the person – should do.) Even if frontal lobe patients have the knowledge of what they want to achieve, their execution of actions to reach their goal can be inappropriate or poorly judged. Many elements of Sue's decision-making and disinhibited interactions with people were consistent with such a syndrome.

Some researchers, such as the neurologist Antonio Damasio, have also demonstrated that patients with frontal lobe damage may not show emotional responses. They can also lack the physiological changes that occur with emotional reactions such as alterations of skin conductance – an index of sweating. This bodily response has classically been used in the polygraph ('lie detector') test and, although not fully reliable for discriminating truthful responses from lies, it has also been used in neuroscience research as a marker of when the body displays an emotional response which might not be apparent overtly. For example, although we can show our emotions when we smile, laugh or frown, many of our emotional responses are covert so other people might not be aware how we feel in a particular situation.

People with frontal dysfunction can lack both facial expressions of emotion and the associated physiological, bodily changes that occur in emotional situations. Damasio proposed that the behavioural changes in these individuals result from their inability to use emotional responses – which they have learned from previous experiences – to help guide their future behaviour.[9]

When presented with moral dilemmas, some patients with frontal damage are also seemingly indifferent to the emotional content of scenarios. For example, in the classic 'Trolley Problem', people are asked to imagine a runaway trolley heading down the tracks towards five workmen who will be killed if the trolley keeps going. They are told that they are standing on a footbridge over the tracks, between the approaching trolley and the five workmen. Next to them on the footbridge is a stranger, who happens to be very large. The only way to save the lives of the workmen is to push this stranger off the bridge and onto the tracks below, where his large body will stop the trolley. He will die if this occurs, but the five workmen will be saved. Would you push the stranger off under these circumstances?

This is clearly an emotional scenario, but if we were considering it from a very rational, utilitarian perspective (i.e., What would be for the greater good?), we might say that it would be worth sacrificing one person instead of five. However, most healthy people balk at endorsing such a course of action and say that they would not do it. It is presumed that this is because it would entail inflicting harm on someone who is blameless and that this is morally unacceptable. But some patients with frontal damage more often choose the utilitarian response, electing to throw the unknown stranger off the bridge.[10] They seem to lack the emotional response to moral dilemmas that normally prevents most of us from selecting this course of action.

Damasio also found that in rare patients who acquired frontal damage when they were babies (e.g., following a road traffic accident or a brain tumour that needed to be operated on), behaviour can be so antisocial as they grow into adults that it emulates that of psychopaths.[11] They can show a pervasive disregard for social and moral standards, may behave consistently in an irresponsible manner and lack remorse. They seem not to have learned conventions of social behaviour and, as a result, they have very few friends.

If people acquire a frontal disorder in adulthood, they are usually not quite so extreme in their behaviour, but they can act inappropriately in social contexts. One important feature of their interactions with other people is an impaired ability to see things from someone else's perspective. They may lack what has been called a 'Theory of Mind', or the capacity to understand other people's mental states – their intentions, motivations, emotions or thoughts.[12] Without such an appreciation and ability to control responses, they can blurt out the first thing they think about, regardless of how this might be received by someone else. This also seemed consistent with Sue's behaviour. Her relationships with her family, friends and neighbours had all been scarred by her corrosive manner.

But what was causing her frontal dysfunction? Sometimes people develop behavioural changes like Sue's in their fifties or sixties as a result of a psychiatric condition that comes on without any previous similar history, but this is unusual. Because her symptoms had progressively worsened over two years, it also remained a possibility that they were due to a slowly growing frontal tumour (such as the meningioma that patient EVR was found to have), so I asked for her MRI scan and neuropsychology to be performed relatively quickly.

~ o ~

Sue and Alan came back to see me two weeks later. This time, Alan had persuaded her not to wear the cowgirl outfit, but she arrived instead in shorts and sandals. It was January and freezing cold. As she entered, she pointed to an article in the newspaper she was carrying.

'You see this, doctor? Just as I said to Alan a few months ago, Angela Merkel's decision to let all those Syrian refugees into Germany is coming unstuck. Some of them are being arrested for touching up young women in Cologne on New Year's Eve. Disgusting! She should never have let any of them in. It's just the start of a huge problem.'

I knew better than to argue. The backlash had begun against the Syrian refugees who had been welcomed into Germany only the previous year (over four years since the beginning of the civil war in Syria). Judgements were being made – and voiced – on the basis of the criminal actions of a few of them.

'What, you don't have an opinion? All quiet, are we, doctor?'

'Please do sit down. Let's discuss your scan and neuropsychology test results, shall we?'

'This is far more important than my scan, doctor. This Syrian refugee stuff is a crucial moment in history. You mark my words.'

Rather presciently, Sue proved to be right. But it was not the appropriate time to start a discussion on how immigration from Syria might have a profound impact upon Europe, which it subsequently did. I brought up her brain scan on the computer monitor and took them both through the images, describing what they showed. The scan had not revealed a tumour, I explained, but it wasn't normal either.

'Well, I'd hate to be normal.' Sue smirked. 'Nobody wants to be average, doctor, do they?' She drummed her fingers on the desk.

'I suppose not. You see here, in the front of the brain and also to some extent the temporal lobes on the side? They look reduced in volume compared to other parts of the brain. They seem to have shrunk.'

'And what might be the cause of that?' asked Alan.

'The most likely cause is an illness called behavioural variant frontotemporal dementia which . . .'

'Dementia! I'm not demented, doctor,' Sue interrupted. 'I've got a great memory, haven't I, Alan?'

Alan nodded. I continued.

'There are many types of dementia, Mrs Ryland. People with behavioural variant frontotemporal dementia can have very good memories, but their behaviour can change in the way that seems to have happened in you. They can start to act differently.'

Alan looked shocked as he said, 'So, if the front part of the brain shrinks people can behave like Sue?'

I nodded. 'I'm afraid so.'

'What causes it, then?' he asked.

'Sometimes there's a genetic cause, but that seems unlikely in Mrs Ryland, given there isn't anyone else in the family who has been similarly affected. It seems that abnormal proteins can be deposited in the brain and they can cause nerve cells not to function properly.'

'Really? Proteins in the brain can do that, doctor?' Sue asked with irony, suddenly curious but nevertheless very agitated, her eyes wide open. 'Proteins in the brain can make you behave differently, can they? I've never heard such nonsense,' she fumed.

'I'm afraid so, and your neuropsychology test results would also be consistent with dysfunction of the frontal parts of the brain.'

The tests had confirmed that she was impaired on those cognitive tests that require people to suppress their immediate response; or switch from one feature to another, such as switching between letters and numbers when searching for items; or multitasking. As we had found in clinic, Sue's episodic and semantic memory seemed intact, and similarly her visuospatial functions were unimpaired. There was no evidence of limb apraxia. Our neuropsychologist had also performed some tests of 'moral cognition' (such as the Trolley Problem),[10] and these had revealed that Sue was far more likely to choose the more logical, utilitarian course of action, even if that would mean directly causing physical harm to someone. Intriguingly, like other patients with behavioural variant frontotemporal dementia, she had a good sense of what was right and wrong, but when it came to making decisions where there were moral dilemmas, the emotional content of those scenarios did not affect her decision-making.[13]

'Oh well, there we are, Alan. You're going to have to look after me from now on. I want the full pampering treatment. Thank you very much, doctor.' And with that she headed for the door. 'Alan, are you coming?'

'Perhaps we might talk about treatment before you go?'

'Treatment? I don't need treatment, doctor. I'm being myself no matter what your scans show. Remember, I'm a liberated woman,' she concluded as she slammed the door after her.

Sue's responses were not a complete surprise. I had anticipated it wouldn't be an easy consultation. Alan stayed behind.

'Is there really some kind of treatment she could try?' he enquired hopefully.

'There isn't a cure but there is something that might help to reduce the agitation and the aggressive, impulsive behaviour. It's called trazodone.'

'And how does that work?'

'We're not quite sure. It was developed for depression, but it's also been shown to improve some of the behavioural symptoms in people with frontotemporal dementia. It changes the balance of chemicals in the brain.'

'Might it stop her getting worse?'

'I'm afraid this is a progressive condition, as you've seen over the last couple of years, but it might help with some of the behavioural issues, at least for a time.'

'I'd like her to try it, please, doctor.'

'Of course, but we first need to get her to accept that there is something wrong and that the medication might be useful for her.'

'I'll talk with her sister and get her to come over,' Alan said. 'Sue might listen to her and give it a go. She's the only one in her family who is still talking to her.' So, I wrote a prescription for trazodone and wondered whether Sue would ever contemplate taking the drug. 'I can call her to explain about the drug if she is interested to try it.'

~ o ~

As I walked across the Square that afternoon, I could not help but reflect on my second encounter with Sue. She was difficult to forget. The sun hung low in the sky, making the elongated shadows of the bare trees stretch across the expanse of grey grass. Winter was in full swing, and Sue had arrived at the clinic in summer apparel, oblivious to her surroundings or what people thought. I smiled at the barista in his coffee van. He was still there. I wondered if he had also had the pleasure of a reacquaintance with Sue that day.

Behavioural variant frontotemporal dementia (abbreviated by

neurologists to bvFTD) accounts for between ten and twenty per cent of dementia cases, but it is often an underdiagnosed condition. Because behavioural changes are often the most prominent signs, patients may be referred in the first instance to a psychiatrist. At that stage, brain scans, if they are performed at all, may appear to be normal, so a neurodegenerative condition might be ruled out. A definitive diagnosis may then take many years to establish. When it is made, it is most often given to people who are between forty-five and sixty-five years old, but recent research reveals that the prevalence of bvFTD increases in even older people.

The characteristic features of the illness include a host of different symptoms: impulsive or disinhibited behaviour, loss of empathy, lack of motivation, change in dietary preference (often with the development of a 'sweet tooth'), and lack of insight, with relative preservation of memory, visuospatial function and praxis. Sue showed many of these attributes. Her story even included an incidence of shoplifting that Alan had managed to scupper, but some patients with bvFTD can become involved in far more dangerous criminal activities and antisocial behaviour. More than a third of individuals with the diagnosis commit crimes ranging from traffic violations, trespassing and theft through to indecent behaviour, wilful damage of property, burglary, physical assault and sexual harassment.[14] These and other transgressions of social norms inevitably place huge pressures on their family members.

Their predicament is compounded by the indifference and lack of empathy shown by many individuals with bvFTD. There is evidence of both a loss of *emotional empathy* and *cognitive empathy* in the condition. Emotional empathy is the capacity to have emotional reactions to how other people are feeling, sharing a 'fellow feeling' with another person. Cognitive empathy, on the other hand, is a term used to capture the ability to see a situation from somebody else's perspective and understand how they feel. The loss of both types of empathy has been related to atrophy of parts of the frontal and right temporal lobe.[15]

A common observation of family members is that bvFTD patients are emotionally detached, 'cold' and show little eye contact. They may make hurtful or insensitive comments, mock a person's appearance or show disregard for someone's pain or distress. Inevitably, this can lead to their exclusion from their social circles and, eventually, even from their families. They become isolated. It sounded as if within two years Sue was a long way down the road to becoming socially excluded.

Would trazodone make any difference? The drug was originally developed as an antidepressant. It works differently to modern SSRI drugs, which selectively inhibit the re-uptake of serotonin, a neurotransmitter, at synapses (the connections between neurons). In this way, SSRIs act to increase the level of serotonin within the brain. Trazodone's major effects are thought to rely on blocking the action of a particular receptor for serotonin, thereby suppressing its activation. Overall, trazodone is not only effective in treating depression but can also improve some of the behavioural symptoms of dementia, including the irritability, agitation and dietary changes that occur in bvFTD.[16] Recently, the Cambridge neurologist and neuroscientist Giovanna Mallucci has provided evidence that trazodone may, at the right doses, also have a protective effect on the brain across many neurodegenerative conditions by acting on an important protein-regulation pathway within neurons.[17] I wasn't sure, though, whether we'd see any change in Sue, or even if she'd take the medication.

~ o ~

Two months later I received an email from Alan. Would it be possible for me to see Sue earlier than planned? He would really appreciate the opportunity to discuss a few things with me. That did not sound promising, but I overbooked my next clinic list to see if I might be able to help. On this occasion, I was surprised. To begin with, Sue was dressed normally, dare one say appropriately, for a clinic appointment. I felt so conformist for even thinking in this way. She had come in a warm coat over a smart dark blue trouser suit. Alan, too, looked less put upon than on our previous meetings.

'Hello, doctor, and how are you?' asked Sue.

She'd never asked me that before, so I wasn't sure where this conversational gambit was going to go, but I responded in traditional British style.

'I'm fine, Mrs Ryland, and what about you?'

'Well, I was never feeling bad, as I told you.' She smiled.

'That's good to hear. Did you try the trazodone, by the way?'

'That's really one reason we're here,' Alan explained.

I waited.

'Sue was persuaded by her sister and you – thank you for the call – to start the medication . . .'

'I do listen to Polly. I know that if anyone has my interests at heart, she does,' Sue interrupted, drumming her fingers once again on the desk. It didn't look like a great deal had changed.

'And we have been able to increase the dose as you suggested,' Alan said, completing his sentence.

'And?' I asked.

'And we've come early to see if we could up the dose again,' he said.

'Because?' I looked at them both enquiringly.

'Because I think it has done some good. Sue's far less, how shall we say, outspoken and now she gives people an opportunity to explain their point of view. She's less likely to get angry or agitated with me or members of the family. So that's been really good, but obviously it hasn't changed everything, as you can sort of see for yourself.'

I was pleasantly surprised.

'Would you agree?' I asked Sue.

'Well, as I said, I felt fine anyway, but Polly explained that I might be getting a little carried away with myself at times, perhaps arguing too much, so if it's working and can help me keep in touch with my family, I'm happy to continue it. So long as you're not slowly poisoning me, doctor!' She laughed, jabbing her index finger in my direction.

I smiled awkwardly. 'No, we're not doing that, but have you noticed any side effects? Are you feeling drowsy during the day, for example?'

'Not that I've noticed. I do sleep better, but that's all I've been aware of.'

'The family's noticed a big difference in how Sue's behaving, doctor. Obviously, they were shocked by the diagnosis, but now we have an explanation of why she's acting this way, everyone's far more understanding. So, thanks for fitting us in today. Would we be able to try a higher dose? I've read we could go up further.'

'Of course, we can try slowly increasing the dose, but did you say you wanted to discuss something else?'

'Sue, I just want to talk to the doctor on his own. Would that be alright?' Alan asked turning to Sue.

'Of course, darling, you ask him what you like,' she responded and gently walked out of the consulting room without a fuss.

I nodded my head, impressed. 'I see what you mean. There has been some improvement.'

'Yes, she's just not quite so wild or aggressive anymore. Thank you for trying the trazodone. It has been really helpful. I wanted to ask you how this will progress. How long have we got and how will she be over the next few years?'

'We see a lot of variation,' I told Alan. 'Some people progress very slowly; others are more rapid. It's difficult to predict, but we'll get a sense of how things are going by assessing her over the next few years.'

'I'm asking partly because the family would like to plan ahead if possible, and partly because we need to decide about whether this is the right time to obtain lasting power of attorney,' Alan explained.

In the UK, a person is deemed to have mental capacity to make decisions until otherwise proven and they need to have such capacity to give lasting power of attorney for their property and affairs, or for their personal welfare, to someone else. But making an assessment of someone's capacity is not always straight-forward. Even if they perform well on a cognitive screening test, as Sue had done, they may nevertheless not have capacity. Formally, to have capacity a person must be able to understand information relevant to making a decision, retain it for the time required to make the decision, weigh up or evaluate the infor-mation to make the decision and be able to communicate what they decide. Sue's ability to evaluate information to make deci-sions had clearly not been good.

'I see. Look, as we seem to be getting some success with trazodone, I think it would be best if we assess her formally with respect to capacity after a few months on the higher dose,' I suggested. 'She might receive some more benefits and may perhaps show better decision-making.'

'Alright, doctor, that seems like a good idea. It's just that I don't want to get to a situation where she couldn't make the decision about lasting power of attorney herself,' Alan responded.

'Have you discussed it with her?'

'Not yet.'

'I think it would be good to start that process at home, gently of course. Perhaps you might do this with Polly since she seems to listen to her?'

I knew, however, that it was going to be a big ask, because

Sue really didn't seem like she would be interested in handing over control on decisions about her finances to someone else, no matter how much they might have her best interests at heart. I brought Sue back into the consulting room.

'All done, Alan?' she asked, smiling mischievously. 'Said what you needed to?'

'He's only trying do what's in your best interests and . . .' I began.

'I know, doctor, and so are you. It's just that I see things differently sometimes. Not today, you'll both be relieved to hear!' She laughed. 'You're safe, for today, Alan.'

'That's a relief to hear,' I said, far too quickly for comfort.

'Who said you don't have a sense of humour, doctor! Did you hear that, Alan? He could be a comedian, this guy.'

A broad smile crossed Alan's face as he stood up to take his leave. Sue, still giggling, flicked his thinning hair playfully and threaded her arm through his. He turned back and raised his hand to say goodbye as he left, pulled along by his wife.

It had been very encouraging to see that trazodone had made a positive impact on Sue's erratic behaviour. With higher doses, she continued to improve. She became less agitated. Some people even enjoyed her approach to plain speaking in conversation. When it was not accompanied by aggressive comments, Sue could seem extremely charismatic and funny. However, she never did consent to power of attorney. She was not going to budge on that, regardless of what anyone said. Inevitably, several years on, we reached a point when even trazodone could not hold back the impact of a progressive neurodegenerative disease on her behaviour.

Of all the individuals described in this book, Sue might be considered the one with the most dramatic change in her 'self' and identity. The alterations that occur in behaviour and personality in people with bvFTD are among some of the most dramatic that are observed in neurological patients. The disease can cause enormous distress to families, friends and even to their doctors, largely because people with the condition repeatedly transgress the boundaries of what is considered acceptable behaviour within society. They break the rules, but occasionally their directness can also be highly appreciated.

7

Is that your hand or mine?

Spring in Ealing, a borough located on the western outskirts of London, was not going well that year. The frosts had suddenly reappeared when least expected, catching the emerging shoots and flower buds unexpectedly off guard in early April. As a result, where the locals were expecting to see a panoply of colour to announce the beginning of warm weather, there was only disappointment. This is how Anna had started her conversation with me in the consulting room when I had asked if she had come far. It was an unconventional answer by anyone's standards, but she seemed nervous and, as a consequence, her response to a simple question had been extremely long.

Apparently, she did not have a garden herself, she went on, but her parents were proud owners of one, and she would often help her mother with her flower beds, but this year had been calamitous. Her expectations for the summer were extremely low. And before I asked about the origin of her surname, yes, her family was originally from Poland where she too had been born, but they had moved to the UK when she was a teenager after Poland entered the European Union in 2004. Like many migrant communities, the Poles had congregated in certain parts of the country and Ealing, as I probably knew, was one of them. Anna had grown up there and, if anything, she now felt more British than Polish.

Yet it had not always been that way. When she was thirteen, she had been walking through a local park chatting in Polish to a friend on her mobile phone. Apparently, this was sufficient provocation for a group of young men that she was passing to start asking her where she thought she was living. 'This is Ealing and we speak English here!' they had screamed at her. When she looked perplexed, they had assaulted her physically, leaving her unconscious at the side of the path.

Fortunately, someone had witnessed the incident and she was taken quickly by ambulance to the local hospital. However, it soon became evident that there had been bleeding within the skull as a result of the head injury that Anna had sustained, so she had been transferred urgently to the nearest regional neuro-surgery unit at Charing Cross Hospital. There she had required an emergency burr hole to be drilled into the skull. This drained the underlying blood and relieved the pressure on the brain below. Or at least, that is how it had been explained to Anna. She was no expert on the subject, she freely acknowledged.

All she knew was that, thankfully, the operation had been a success and surprisingly she had walked out of the hospital only a few days later. Of course, there were follow-up assessments but none now for many years. She seemed to have made an excellent recovery and all had gone well over the ensuing years. She left school with good grades and now, in her early twenties, she was working as a senior administrator in a bank in central London, where she commuted to every day. Did I want any more background?

I had learned an awful lot about Anna, far more than I found out about most patients within the first few minutes of meeting them.

'My goodness, I'm very glad to hear that you did so well after the surgery. So, what brings you to see me?' I asked.

She looked around the consulting room, averting her gaze.

'I'm not sure how to put it, or even if you're the right person to see, to be honest.' Anna responded with a shy smile, pushing her straight auburn hair back over her narrow, freckled face.

'Tell me what the problem is and I can see if we can help.' I offered her a reassuring look.

She studied me carefully, fixing me with her wide, dark hazel eyes, as if contemplating whether I might be someone she could confide in. Eventually she spoke, now in a much softer, slightly tremulous voice.

'It's a bizarre thing to explain to someone, but my right arm and leg have been acting kind of strangely.'

'In what way?'

She hesitated. 'I can be reading in bed and suddenly realise that I don't know where my right hand is, but if I look at it, it will be there. The same can happen with my right leg. I can be sitting on the bus and become unaware where it is. Before I know it, someone has tripped over my foot because it is in the middle of the aisle.'

'How long have you been experiencing these symptoms?'

'Probably for the last six months, but I hadn't really thought too much of them. I just thought I wasn't paying enough attention and that it wasn't anything to worry about. But I'm noticing it more now.' She nodded, apparently gaining more confidence in explaining the reason she had come to see me.

'And it's only your right arm and leg?'

'Yes. I've never experienced anything like that on my left side.'

'Tell me a bit more about when you become aware of these symptoms. What are you doing?'

'It can happen anywhere if I'm busy doing something.' Her voice was stronger now. 'Suddenly I realise that I don't know where my right hand or foot is. But, like I said, if I look at them, there isn't a problem. If I'm not looking at them . . . well, I kind of lose them. I am getting a bit worried about it because it's now also affecting my dancing.'

'Dancing?'

'Yes, I love to dance.' She smiled. 'I joined a club a year ago. We do all sorts of ballroom dances. It's my big hobby. I absolutely adore dancing and spend any free time I have practising. I'm hoping to enter competitions at some stage, but that's not the important thing. I just really enjoy it.' She beamed. It clearly did mean a lot to her.

'That's good to hear,' I said, 'but how is your dancing being affected?'

Anna's face turned a vivid grapefruit pink. Her lips gripped together in obvious discomfort. Whatever it was, her demeanour had suddenly changed.

Eventually she blurted out, 'I know dancing isn't a serious thing but it's where this first happened.'

I nodded, not really knowing what she was about to say, but giving her the space to articulate it without interruption.

'Because I don't know where my right hand and leg are when I'm dancing, it can be awkward with my dance partner . . .' She trailed off, now pursing her lips even more tightly.

'In what way?' I asked gently.

Anna's face contorted into a grimace as she shifted uneasily on her chair.

'Well, sometimes my right leg will wrap itself around his leg and bring us far closer than we should be, if you see what I mean.'

'I see, and what's happened then?' I asked.

'Well, it's really embarrassing. It's one thing not to know the dance steps and mess things up. But it is completely different if it looks like you're coming on to a man on the floor, in front of everyone. One of my dance partners, who I really like, is now reluctant to dance with me anymore. God knows what he thinks of me. But there's another one who's become super keen because he's got the impression that I'm interested in him. I'm not, but either way, it's awkward.' She looked disconsolate, her eyes pinned to the floor in apparent shame. A silence fell over the room.

'That must be very difficult,' I said. 'How long does it take for you to "lose" awareness of where your arm or leg is if you don't look at them?'

She looked up, grateful to be asked a question.

'I'm not really sure.'

'Shall we try now?' I asked gently.

'Alright, but what do you want me to do?' She glanced at me, quizzically.

'I'd like you to look at your right hand. Good, and now look away. Tell me when you start to feel that you don't know where it is.'

It took about twenty seconds before Anna said that her right arm began to fade from her awareness. But the moment she looked at it again, it 'reappeared'. The same also occurred with her right foot when we tested that.

Perhaps she had some unusual sensory loss I thought, but it was odd that her symptoms were confined to only the right arm and leg. I performed the neurological examination – including simple sensory testing of touch, temperature, vibration or proprioception (where she felt her limbs were) – but found nothing abnormal. The only difficulty Anna had was working out what object I placed into her right hand when her eyes were shut. Even without vision, by moving our fingers over an object (a process sometimes referred to as active touch), most of us can normally differentiate between objects such as various types of coin, or a key, or an eraser. But although Anna could do this easily enough when I placed those items in her left hand,

211

she was very unsure what the objects were when using her right hand.

Over a century ago, neurologists in Europe had become aware that such an impairment in recognising objects by touch was usually associated with damage to the parietal lobe on the side of the brain opposite the affected hand. Later, in Montreal, the neurosurgeon Wilder Penfield developed a technique to directly, electrically stimulate the exposed surface of the cerebral cortex of awake patients who were being operated upon for epilepsy.[1] By using this method he was able to map out both the primary sensory and motor representations in the human brain, in order to avoid damaging these regions when he performed the surgery. His painstaking work revealed how sensations from different parts of the body were located in a systematic, topographical manner in the anterior part of the opposite parietal cortex (left parietal cortex for the right side of the body, and vice versa). This is sometimes referred to as a homunculus (a little version of a body). Penfield found that more cortex is devoted to those parts of the body with the greatest sensitivity to touch, such as the face, tongue and hand (**Fig. 14**). These parts of the homunculus are allocated more space in the cortical representation.

When this part of the parietal cortex is damaged, appreciation of sensory information from the opposite limbs can be impaired. In Anna's case, though, it was not simple sensation such as touch that was lost in her right hand, but an inability to integrate the information from touch receptors in the skin with the moving position of the fingers. In the 1950s, Derek Denny-Brown, who worked at Queen Square before becoming Professor of Neurology at Harvard, proposed that the function of the parietal cortex goes beyond simply registering the sensations on the skin. By spatially 'summating' different types of tactile information as the fingers move over an object, a critical function of the parietal cortex, he argued, is recognition of the form of an object –

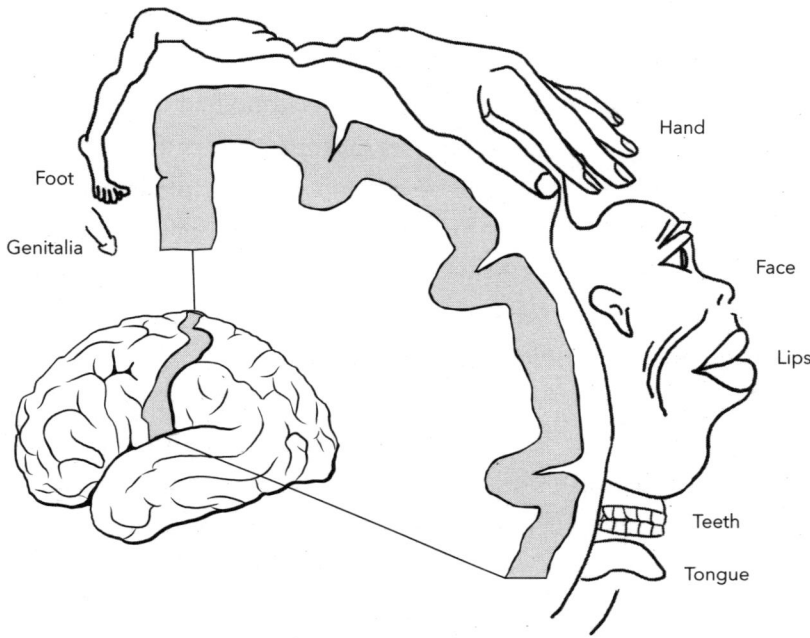

Foot

Genitalia

Hand

Face

Lips

Teeth

Tongue

Fig. 14 | Representation of touch in the parietal cortex. A map of the opposite side of the body is laid out in a topographical fashion (sometimes referred to as a 'homunculus') in the anterior parietal lobe, with more space in the cortex devoted to those areas of the body which have the greatest sensitivity to touch.

which he termed 'morphosynthesis'. When this is impaired, a patient might suffer from 'amorphosynthesis', or an inability to perceive the shape of an object placed in the hand opposite to the parietal damage.[2] 'Which side of the skull did you have your burr hold drilled?' I asked Anna.

'This one. This is where they did the operation.' Anna pointed to the left side of her head as she pulled away her hair to show me an indentation the size of a coin, high up on the back of her skull over her parietal lobe. It was difficult to believe that it was a coincidence that she was now experiencing a loss of awareness of her right arm and leg when these limbs would correspond precisely to the location of her previous surgery.

I was contemplating the possible causes of this phenomenon, when Anna softly asked, 'So you believe me?'

213

I was taken a little aback. I had not thought I had given her any reason to think otherwise, but clearly she had been worried that her story might have sounded far-fetched.

'I have to say that honestly I've never had a patient who has come to me with quite the symptoms you are experiencing – but, yes, I believe you. I think we have to look into this a bit more.'

'Thank you,' she said with obvious relief. 'I didn't know whether anyone would take me seriously.'

~ o ~

I took the tube home that evening. The Piccadilly line train was, as usual, very crowded, jammed with commuters from central London making their way home westwards. As I stood uncomfortably holding a handrail, pinioned between several passengers, an experienced traveller in front of me was nonchalantly managing to read his newspaper despite being squeezed against the doors by the occupants of the carriage. Even though he was being jostled erratically as the train hurtled forwards, he had folded his paper beautifully into one eighth its normal size so that he might easily read each section without disturbing anyone. As he turned it over, I could see a report of how a hospital in the besieged city of Aleppo had been struck by a missile, most likely from a Syrian aircraft, an action which was presumed to be part of a strategy of targeting hospitals in the war. The doctors had been dealing with mounting casualties, including children with head injuries, but now the entire paediatric unit was destroyed.

It was impossible to fathom how people might be behaving in this way, but I knew you did not have to go to Syria to find thuggish brutality. It obviously existed closer to home too, I reflected, as my thoughts returned to Anna. She had been lucky to get to a neurosurgical team so quickly when she had been assaulted simply because she wasn't speaking English. The burr hole that had been drilled was indeed the fastest way to drain the blood from under the skull and relieve the pressure on the underlying brain. Trepanation (making a hole in the head) is exactly the procedure that Marcel Proust had been keen to have performed on himself when he was convinced that he was having a stroke (as discussed in the Preface).

This kind of operation has a long history, dating back over eight thousand years to the Mesolithic era, and is perhaps the earliest surgical procedure for which there is evidence from archaeological digs.[3] Trepanation may have been conducted to let out evil spirits or perhaps to treat neurological conditions, including epilepsy, headache and wounds to the head. Whatever the reason for performing it, it seems to have been a widespread practice in cultures around the world.

But what could be happening to Anna now, so many years after her operation had been performed? Complications of burr hole surgery can occur soon after the operation and include excessive bleeding, which might be difficult to stop, and infection at the surgical site. Sometimes it might be possible that irritation of the underlying brain substance by the surgery or the blood might lead to focal seizures caused by epileptic discharges of the neurons in the brain underlying the burr hole. However, these would usually declare themselves much sooner than the more recent symptoms that had developed in Anna. In addition, focal sensory seizures that might occur as irritation of the parietal cortex usually manifest as tingling or numbness in the opposite hand, not loss of awareness of it. Nor do such seizures stop simply by looking at the hand, so this seemed a very unlikely explanation for Anna's descriptions of the transient loss of her limbs.

The tube rolled into Green Park underground station. Thankfully, many people were getting off and far fewer were boarding here, so there was some respite from the sweaty, crammed pressure of bodies. My impressive co-passenger had turned over his newspaper yet again, to reveal another story, this time about the 'Vote Leave' campaign for the forthcoming Brexit referendum on whether or not the UK should stay in the European Union, scheduled to occur in June. The headline was about the claim that if we left the EU, the National Health Service would be set to receive £350 million every week from the money saved. This message had just started to be used on posters, but would later be emblazoned on a notorious red campaign bus. Even then, in April, the UK Statistics Authority was calling it out as misleading. It wouldn't matter, I thought. People who want to leave the EU were convinced by the propaganda on this and also by the pledge that immigration would be more tightly controlled. The carriage doors closed, and the tube slowly edged out of the station.

~ o ~

I had requested an MRI scan of Anna's brain to see if there might be any structural cause for her symptoms, so we would discover soon enough if there was something related to her previous surgery that was to blame. Two days later, far earlier than I had anticipated, I heard about Anna again, but this time via a message left by the police with my secretary. They had detained her following an event on a bus. Would I be able to call them to discuss the situation? I wondered what might have happened and decided to respond immediately. A duty sergeant answered the phone.

'Right, sir, so you know Miss Anna Kowalska?' he asked in a distinctly ponderous manner.

'Well, I saw her earlier this week when she came to my neurology clinic.'

'I see, sir. She's got herself into a little scrape and she gave us your name as an expert who might be able to help.'

216

'Of course, I'll try to help if I can, sergeant. What seems to be the problem?'

'We received a call from a very distressed young woman who was sitting next to your Anna on the bus . . .'

'She's not "my Anna", sergeant,' I interrupted quickly.

'So sorry, not a good turn of phrase, sir. What I meant to say is that she was sitting next to your patient on the bus when she felt Miss Kowalska's hand touching her thigh. Let me see, yes, she actually described it as "groping her thigh", sir.'

'Which hand was it, sergeant?'

'Does it matter, sir?'

'Yes, it does,' I responded curtly.

'Well, looking through the notes, I think it must have been your patient's right hand.'

'I see. There might be an explanation for this, sergeant.'

'Really, sir?'

'Yes,' I said. 'Anna is actually being investigated for symptoms where she loses awareness of the right arm.'

'Loses awareness, sir?'

'She doesn't know where her hand is.'

'And that's a neurological symptom?'

'It can be,' I told him. 'She had a hole drilled into the left side of her skull and it is possible that she is experiencing an after-effect of this. We're waiting for a brain scan to see if we can do anything about this.'

'A hole drilled into her skull, you say?'

'Yes, she needed urgent surgery because she had been assaulted. This happened a few years ago, but I think she's experiencing some late effects of this.'

'So, you think this touching, shall we call it, isn't her fault. She didn't do it?'

'Does she say that she did do it?'

217

'Well, that's the problem, sir. She's been very apologetic to the woman who lodged the complaint, so it sounds like she did do it.'

'Her hand might have touched the passenger next to her, but she might not have been aware that it had done so.'

'Seems very strange to me, sir. She did it, but she wasn't aware of groping the woman next to her?'

I sighed. This was always going to be difficult to explain.

'The brain is complicated, sergeant.'

'I'm sure it is, sir, and you should know, no doubt.'

'It is possible for some patients to be unaware of what their hand might be doing.'

'And you're willing for us to put in the records what you have just said about her, what's the term, "losing awareness" of her right hand?'

'Yes, of course,' I replied.

'Then we can let her go with a caution, but we will have to come back to you once our investigations are complete.'

'Thank you, sergeant. May I speak to the patient?'

As I waited to speak to her, I thought how Anna's hand was clearly now becoming a problem, transgressing people's personal space outside the context of her dance club.

When she eventually came to the phone, she was very distraught. She evidently felt so ashamed by what had happened. I tried to reassure her, but she started to sob breathlessly.

'What happened, Anna?'

'It was horrible. I was daydreaming on the bus when the woman next to me started to shout. I looked over towards her and I could see that she was looking down at her legs. I glanced down and thought she was clasping her thigh because she had seen something frightening, but it wasn't her hand. It was mine that was squeezing her thigh. I mean, it wasn't just brushing it. It was actually grabbing her thigh . . . She had a short dress on, which made it even worse. This has never happened to me before.'

'So, your right hand was holding her thigh, but you weren't aware of it?'

'No, I had no idea what it was doing. It's like it had a mind of its own.'

I paused. 'Anna, Is there a family member who might take you home?' I asked.

'Oh no, I don't want anyone to know I'm at the police station. Don't worry, doctor, I'll make my own way back.' There was another long silence. 'Thank you for calling, though. The police didn't really believe me when I started to explain, but they're a little more understanding now. It's been such an awful experience being here.'

~ o ~

Three weeks later, Anna's scan report came through in my email inbox. It was not normal. I asked my secretary to call her and arrange for her to come in on my next clinic day.

'How have you been?'

'I'm desperately making sure that I keep looking at my right hand and foot to make sure that I don't lose them when I'm in a public place. I was so embarrassed with the event on the bus.'

'Have there been any issues at work?'

'Oh no, the police didn't inform them.'

'That's not what I meant. Has your arm or leg caused any problems at work?'

'No, I really try very hard to pay attention now when I'm with other people. What did the scan show?' She looked at me anxiously.

'Well, that's why I brought you back to see me earlier than I'd planned.'

She gazed at me quizzically. 'And?'

'I'm going to discuss the scans with you in one minute but might I first re-examine your hands again? If you could rest them palms facing upwards on the desk. That's right and just keep them still, do nothing.'

I brought my empty coffee cup close to touch each of Anna's hands. There was no response.

'Now shut your eyes for me and again just keep your hands still – do nothing.'

I waited for twenty seconds and then touched her left hand with my cup. Again, nothing. Then I touched her right hand with the cup and this time suddenly, as soon as I did, it began to grasp at the cup, holding on so tightly that I could not prise it back.

'You can open your eyes now, Anna.'

She gasped and released the cup immediately.

'I didn't know that I was holding on to this.'

'No, I think you had lost awareness of your right hand, but it responded to the touch in a reflex manner by grabbing hold of it. I think this is probably what happened on the bus.'

Anna brought her hands to her mouth and shook her head.

'Anna, your scan shows a cyst over the brain, where you were hit – where the hole was made by the neurosurgeons.'

'A cyst? Do you mean a tumour?' She asked, clearly horrified at that prospect.

'No, no. It really doesn't have the appearance of a tumour. It is a fluid-filled cyst. We call it an arachnoid cyst.'

'A what? It sounds terrible.'

'An arachnoid cyst. They're quite common in children and may not cause any symptoms at all. But they can also develop as a result of a head injury. They lie between the brain and the arachnoid membrane, which is a protective layer around the brain, and they contain the fluid that normally bathes the brain. They're not malignant. I had been worried that something had happened around the burr hole in your skull, but I think you developed a cyst after your head injury, and I suspect that it's slowly grown over time.'

'Slowly grown? But does that mean it will keep growing? And how is it causing this problem with my arm and leg?'

'I can't be sure about the rate it will grow, but the fact that you've developed symptoms only now makes me think it has been growing quite slowly over the years since your head injury. Now it's got to a size where it is pressing on the underlying brain – on the left parietal cortex. This part of the brain interprets the sensations from the right arm and leg. The pressure from the cyst may be preventing it from working normally.'

Anna looked wide-eyed. 'I thought the head injury was behind me, but through all these years it has actually left a mark, a cyst which is growing?'

'I'm afraid so.'

'Is there any medicine I can take to make the cyst shrink?'

'Not really, but you could have the fluid drained.'

'Drained? How can it be drained?'

'The fluid could be stopped from accumulating by putting in a shunt that drains the fluid into your tummy or . . .'

'Into my tummy!'

'It would involve a small tube that goes from the head, under the skin to your abdomen.'

'No, I wouldn't want a tube like that. It sounds horrible.'

'Or you could have the cyst surgically removed.'

'Another operation on my brain? No thank you. That would be too scary.'

'I understand that this is all shocking news for you, Anna, but let me show you the scan so you can understand what I'm talking about.'

'I'm really not sure I want to see the scan, doctor. I thought I had left my head injury behind me a long time ago. I thought I had recovered from the stress and anxiety that created, but now it's all coming back.' She hesitated. 'You see, even months after the operation I dared not walk in that park alone. I was so

221

frightened.' She looked down to the floor. 'And now that fear is coming back.' She paused. 'I'm really sorry. I know you're only trying to do what's best for me, but I don't think I can talk about operations and scans right now.'

'Of course, we don't have to,' I assured her. 'Why don't you go and think about what I've said and we'll see each other in a few months' time?'

She left quietly, despondent and clearly shocked by the news.

~ o ~

Although she did not want to see it, Anna's MRI brain scan had revealed quite a large arachnoid cyst. As I had explained to her, these can occur as the brain develops in children and usually do not require any intervention. In fact, they may only become apparent as an incidental finding on a brain scan performed in later life for a completely different reason. In Anna's case, though, the cyst seemed to have developed following the physical trauma of the head injury, which is recognised to happen in some people. Furthermore, the location of her cyst over the left parietal lobe corresponded well to the strange symptoms she reported involving her awareness of where her right arm and leg were, because this is an area which plays a key role in generating the 'body schema'.

At the turn of the twentieth century, British neurologists Henry Head and Gordon Holmes developed the concept of a representation in the brain that keeps track of the position of our body parts (fingers, hand, arm, leg and foot), as well as all the tactile inputs from the sensory receptors in the skin across these areas. According to Head and Holmes, every new movement we make updates this representation, which they called the body schema, in a continuously dynamic manner.[4]

Modern neuroscientists consider part of the parietal cortex to have a crucial role in creating this body schema. It is now established that sensory information from different types of skin receptors (responsive to light touch, vibration, or temperature), and proprioceptive receptors that signal posture (joint angles), from throughout the opposite side of the body converge in the parietal cortex (left parietal cortex for the right side of the body, and right parietal cortex for the left side of the body). These inputs are essential to producing a representation of where our body parts are in space and what they are touching. Cysts like the one Anna had developed can disrupt such representations.[5]

A body schema is considered to be vital not only for passively tracking what is happening to our limbs but also for planning the consequences of their movement.[6] If I move my hand in such a direction, will it hit a glass in its path or avoid it so that I can pick up the bottle I am aiming to reach? This kind of subconscious planning of movements in the space within arm's reach – referred to as 'peripersonal space'[6] – is crucial in everyday life. But it's not just the movement of our limbs that is important. Whether it is a knife, fork, spoon, screwdriver, pair of pliers or a litter-picker tool, we are able to compute with ease where our hand will place the end of a tool to use it effectively. There is now increasing evidence that tools are also incorporated as an extension of the body schema to allow us to do this. Similar challenges confront the next generation of smart robots – and the scientists who are designing them. Computationally, creating a three-dimensional representation of the body in space is a formidable challenge, given the vast array of information that has to be processed to calculate where body parts are and what they are feeling.

Anna seemed to have a dysfunctional 'body schema', but the impairment appeared to be isolated to the representation of her right arm and leg. She needed to use her sight to keep track of where these limbs were, and if she didn't do so her hand could bizarrely grasp at objects it came into contact with. This involuntary grasping resembles a rare condition called the alien limb syndrome. Patients suffering from this disorder lose voluntary control of a hand which can seemingly act autonomously, grasping nearby objects tightly. They can often be embarrassed by the actions of their wayward hand, which in some cases has included fondling other people or even placing them in choke-holds, without the patient wanting to do this. The character of Dr Strangelove, portrayed by the actor Peter Sellers in the film of that title, is afflicted in a similar way, but goes one step further by attempting to choke his own neck with his hand.

But why should Anna have developed an 'alien hand'? Neurologists recognise that such involuntary grasping can occur for several reasons. In Anna, it seemed as if loss of the representation of her right arm and leg had led to them becoming almost independent. Free from the constraints of a body schema, they seemed to act autonomously, so if her hand was in contact with anything it would attempt to grasp the object. But sight of her hand would return 'possession' of it to her. Unlike Anna, other patients with an alien limb may involuntarily grasp an object near them when they can see or feel it. Again, the hand seems to act reflexively and of its own volition.

For most of us, the very idea that part of our body can act independently of our intentions seems bizarre. Throughout our lives, we are used to ownership of our body: it is a given fact. We have agency over its actions. But when part of our body seems to act autonomously, this calls into question how much possession we really have. Normally, we will our body to act in different ways, integrating the sensory and proprioceptive inputs with the motor commands we send to our muscles, building up a schema of the body in the world. But what would happen if we lost that schema entirely? Anna had not suffered such a complete, catastrophic loss. But some very rare people seemingly have, not from

damage to their brain, but because of loss of proprioceptive information – knowledge about the position of their body parts.

Ian Waterman is such a case. At the age of nineteen, he
suffered a viral illness which triggered a rare autoimmune
disorder. This damaged the cell bodies of the sensory nerves
that carry information about position sense (where his limbs
were positioned in space) and touch to his brain. Ian effectively
became 'deafferented' from the neck down.[7] For months, even
though his motor nerves were intact (so he could theoretically
move his arms and legs), he needed continual nursing care. He
was unable to move because he did not know where his limbs
were or what they were touching. Gradually, he began to make
effective, coordinated movements with his arms, but only if he
looked at his hand as he did so. He had no knowledge of where
his hand was unless he could see it. Ian commented that it
required enormous concentration to make even the simplest of
movements. It took him several weeks to learn how to put
something into his mouth to feed himself, and even longer to
be able to walk. And in the dark he always remained vulnerable,
unable to make any movement safely.

The philosopher Shaun Gallagher and the neurophysiologist
Jonathan Cole – it was the latter who first documented Ian's
unusual sensory loss – argue that he effectively lost his body
schema and therefore had to develop new strategies to move in
the world.[8] For them, the body schema plays a vital role in what
is known as the 'embodied self'. Whereas some thinkers in the
past, going back to the time of Descartes, have argued that the
self can exist without the body, others have been wary of
dismissing the special role our bodies play in defining ourselves.
It is evidently the case that we occupy a spatially delimited object
– our body – which is different from the bodies of other selves.
For many philosophers and psychologists this means we cannot
ignore the fact that our perceptions, actions, memories and all

other cognitive processes emanate from an embodied self.[9] In their view, we cannot understand the self without the body.

As I walked out of my clinic room, I reflected on the fact that Anna had effectively lost part of her body schema and, with it, part of herself: her right arm and leg to be precise. She'd lost awareness of those limbs unless she looked at them – a much milder form of the symptoms that Ian Waterman had suffered, but nevertheless they were having a serious impact on her life.

'The young lady you just saw, doctor, asked to make her next appointment in a year's time rather than four months. Is that alright?' asked the receptionist as I was passing her desk.

'Yes.' I nodded. Evidently, Anna really didn't want to contemplate surgical intervention, which was perhaps understandable, given the mental trauma of her initial assault and all the memories associated with it.

~ o ~

'We have a patient of yours in the Emergency room, doctor,' the voice on the other end of the telephone announced brusquely.

'And you are?' I asked.

'I'm one of the nurses. This Anna Kowalska got blue-lighted in this morning. The ambulance crew were called to her work. They found her having a grand mal seizure, but weren't able to stop it immediately with diazepam. Eventually, with more of the diazepam, the seizure did stop but she was very confused. She's now speaking more coherently and the doctor who saw her asked me to give you a call as he's been bleeped to another emergency.'

'OK. Has she had a brain scan yet?'

'Yes, I believe that doesn't show any change, but it should be available on the computer system if you want to take a look. It's just that we need to make a decision about whether we keep her or discharge her soon. The place is filling up.'

'Alright, it might be easiest if I come to see her. I'll be there soon.'

Anna's CT brain scan showed only the cyst. It didn't seem significantly larger than on her previous scan, which had been performed just a few months ago, but the seizure was an unexpected new development and at the very least she needed to be started on medication to prevent further occurrences of seizures.

I started to walk over towards the Emergency department located at our sister hospital. The low-hanging clouds that November morning were leaden with a deep Prussian blue, over which shades of slate, ash and white swirled and ballooned ominously upward. It was a declaration of intent for which I was ill prepared. As I threaded my way through the archway of Senate House and walked up Malet Street, the deluge commenced. It was violent right from the off, beating down malevolently on the streets with an unrestrained passion. Some pedestrians fortunate enough to have umbrellas had managed to prise these open against the onslaught, but the rest of us were caught in what seemed like a ceaseless fusillade of rain. The water cascaded over me, unrelenting in its fervour, as I decided to run.

By the time I stepped into the Emergency department, I was drenched. One of the porters could not help but smile as I passed by him. 'A little wet out, doctor?' he asked, wryly raising an eyebrow, as rivulets of water coursed down my face. I didn't answer but simply acknowledged his irony with a nod, leaving a trail of small puddles in my wake. 'Perhaps you might find these useful?' he said rushing up with some paper towels, just managing to restrain himself from laughing out loud. I thanked him and, as I mopped my hair, caught sight of my bedraggled face in a mirror by one of the sinks. It wasn't my best professional look, but it couldn't be helped, I told myself.

When I pulled the plastic blue curtain back, Anna was sitting up on a trolley bed. She looked dazed, her face a picture of quiet desperation. She glanced up and was evidently surprised to see me. To her side a small woman with a hood of very dark straight hair and pointy features, almost like a finch, stood perched, her hands firmly holding the rails of the trolley. She looked me up and down with suspicion. I was still dripping.

'Doctor, you work here too?' asked Anna, surprised to see me.

'I got a call and came over.'

'Oh, thank you. Mama, this is the neurology doctor I told you about,' Anna said turning to the woman by her side.

'*This* is the specialist?' her mother asked with her eyebrows raised in disdain.

'I'm sorry for my appearance,' I said as a particularly large drop of water fell from my forehead onto the floor. 'I came over from Queen Square, but it started to rain quite violently on my way. I would shake your hand, Mrs Kowalska, but as you can see it's quite wet.'

'No problem,' Anna's mother said, obviously pleased that I knew – and could pronounce – her surname, but nevertheless still dubious about me. Unlike Anna, she spoke with a strong Polish accent. 'I arrived only half an hour ago but we haven't seen anyone, doctor. Can you tell us what has happened, please?' she asked.

'Of course, I'll try my best,' I responded. 'But perhaps I might ask a few questions of Anna first. How are you?'

'I feel terrible,' she said, holding her head in her hands.

'Do you remember what happened?'

'Up to a point. I was in work, talking to a colleague when I suddenly started feeling weird. My right arm started to tingle and then it started to jerk and I could feel my right leg start to move, and that's all I remember.'

Her descriptions, together with what the nurse had recounted to me, sounded as if she had first experienced a focal seizure affecting the left parietal lobe, which had produced the tingling in her right arm. The abnormal electrical activity must have quickly propagated throughout the brain, leading to loss of consciousness and a generalised epileptic seizure, with shaking of all four limbs.

'It seems like you had a seizure, Anna.'

Her mother shook her head in disbelief. 'You think Anna has epilepsy?' she asked. 'How can that be?'

'Well, I think she has had a seizure, Mrs Kowalska. That doesn't mean she has epilepsy, because strictly speaking to get that diagnosis a person needs to have had two or more seizures, which have occurred more than a day apart. So, Anna's just had a seizure.'

'OK, I see, not epilepsy,' said Anna's mother looking a little more reassured. 'And what has caused this? Do you know? Can you tell us?'

I looked towards Anna to check that she was happy for me to explain further.

'It's alright, doctor, my mother knows about my scan result and what's been going on. You can tell her everything you would tell me,' Anna said.

'Yes, yes. I am her mother. You can tell me, doctor,' her mother added firmly.

'The CT scan you had this morning doesn't show much change in the cyst, Anna, but I think there is little doubt that it's the cause.'

'But how? How does this . . . cyst . . . make a seizure?' queried Anna's mother.

'Well, it can put pressure on the brain underneath it. That would be the most likely cause.'

'Could it happen again?' asked Anna.

'I'm afraid so. That's why we have to start you on some medication to prevent seizures.'

'Do I have to take it for life?' she asked, alarmed.

229

'If the cyst is there, I'm afraid there will always be a risk of seizures.'

'And if it isn't there?' asked her mother.

'There would be less risk and we might be able to stop the medication after a while, but you definitely need something to protect you now. Do you drive, Anna?'

'No, no. I don't drive.'

'Then that's one less issue to worry about,' I reassured her before asking, 'How have you been otherwise since we last met?'

She looked down. 'Not great, doctor. It's been quite hard.'

'What's happened?'

'I've had to stop going to the dance club because . . . because of my problem.'

'Because?'

'Because my hand got me into trouble again.' She shook her head, upset even before she told me.

Her mother put her arm around her.

'I wasn't even dancing. That's the silly thing about it. I was just standing at the bar waiting to order a drink, probably distracted and not thinking about my hand. Suddenly, this woman next to me shouts out, "Do you mind!" I looked towards her and then down to my right hand, which was touching her bottom. I could have died. I apologised and explained I had a condition, but it was clear that other people had seen it. The chairman of the club came over to talk to me. Thankfully, he didn't take it further, but I can't go back there. It's too embarrassing. If it were to happen again, I don't know what I'd do.'

Tears were streaming down Anna's flushed cheeks.

'People, they can be so ignorant. My Anna would never do something so rude,' her mother said as she comforted her daughter. 'It's not her, it's her arm, doctor.'

'The dance club means a lot to you, I know,' I said quietly.

'It's my world, doctor. That's where I've met people, made friends. I mean, I have told some of my girlfriends about the cyst, but they don't really understand, and I just look like a pervert to everyone else.'

'Look, Anna, this cyst is now having a much bigger impact on your life than a few months ago. You've had a seizure today and now you can't go to the dance club which is so important for you. I realise you didn't want to talk about surgery before, but I think it might be good if you at least talked to one of our neurosurgeons, so that they can take you through the options.'

Anna looked down at her hands and nodded. 'OK, I'll talk to them, but it's so hard for me, doctor,' she said as she burst out crying.

'I know. I understand. Let's see what they can do.'

'Doctor, may I have a talk with you?' Anna's mother asked.

'Of course, Mrs Kowalska.'

'I mean in private.'

'Would that be alright with you, Anna?' I asked. She agreed, and so I stepped out of the cubicle and walked with her mother to an interview room, which thankfully was free.

'Look, doctor, Anna is our only daughter. She is a beautiful person. She has gone through a lot when she was a child. After the brain surgery she was so scared that she would be, how do you say, attacked again. It took years for her to get her confidence back. She has a good job and the dancing is the way that she has made new friends. She was happy, very happy.'

'I understand, Mrs Kowalska.'

'But maybe you don't understand, doctor. Now that she feels she cannot go back to the dancing, all she does is work and come home. She is too . . . too ashamed . . . to see anyone. It's not her fault, but this is ruining her life. She is losing her friends because of this crazy cyst.'

I nodded my head. 'I do understand. This is why I think we need Anna to see a neurosurgeon and—'

'She will be so frightened to see a brain surgeon again, doctor. It is like walking back into the past when she had left it so far behind, you know.'

'I think I know. I understand that it will be a real challenge . . . it will be very difficult for Anna to see a surgeon, but I think it would be good for her to find out what they can offer. I promise I'll send her to someone who will be kind and who will be aware of what you are saying.'

'You promise? You really promise?' she asked looking me directly in the eye.

I nodded. 'I promise.'

'Then I shake your hand, doctor. It is not so wet now.' She smiled approvingly.

~ o ~

Most neurosurgeons are optimistic creatures. They have to be. Their line of work necessarily involves taking risks on our most delicate and precious organ. Any unwanted effects of surgery can have devastating consequences for a patient. This is why neurosurgeons are also very careful to consider the risks they will put an individual through. Most importantly, they are always asking themselves whether those risks actually outweigh the potential benefits of surgery. If they don't, they usually won't offer to operate.

Anna had apparently been very reassured by the neurosurgeon who I referred her to. It had surprised her to find that it was a woman who saw her, and even more that she had been sympathetic about her concerns. She hadn't felt rushed, which is always a worry in a busy clinic. Gratifyingly, she had been given time to pose all her questions and she'd also been asked whether there was anything else she wanted to discuss. This was very reassuring for me to hear, as I had primed my neurosurgical colleague about Anna's circumstances and how fragile she felt. It was good to hear that Anna had such a positive experience meeting her.

Apparently, Anna had consented to the operation there and then. She was admitted a couple of weeks later. The surgeon had been able to incise the wall of the cyst, drain the fluid and stitch the cyst lining, so that it would not re-form. There had been no need for a draining shunt to the abdomen, which pleased Anna enormously. She had been discharged two days later and was extremely happy with how things had gone. True, she had a large scar, but her hair was growing back over it and soon it would be hidden.

'I'm so glad it went well, Anna. How are you feeling now?'

'Good. I've gone back to work last week and everyone's been very kind.'

'And you've not experienced any more seizures?'

'No, thank goodness. I was worried what might happen with the stress of the operation, but I've not had any more seizures.'

'And how about your right arm and leg?'

'Hmm. Yes. Well, they're much better behaved too,' she said, beaming.

'In what way?'

'Well, I don't lose them anymore, or at least I don't seem to lose them.'

'Explain what you mean.'

'Normally, if I looked away, I wouldn't know where my right hand and foot were after a little while. I think you showed me that it was about twenty seconds. That seems to be much better now. In fact, I don't seem to lose them at all if I look away, no matter how long it's for, but that's when I'm on my own. I really don't know what it would be like if I was with a group of people or in a public place. I caught a taxi here today because

I didn't want to test what would happen on a bus.' She looked anxious again.

'I see,' I said. 'So, you're worried that it will still happen if, say, you went back to the dance club?'

'Yes, exactly.' She nodded. 'It would be too embarrassing if my hand misbehaved again, doctor.'

'Well, let's try a test, Anna, shall we?' And with those words, I went outside into the corridor where there were several medical students waiting to sit in on clinics, and returned with one of them.

'Anna, this is Sarah, one of our medical students. She likes to dance. I've explained to her your concerns about losing awareness of your right hand and leg when you are on the dance floor. If you're willing to try this small experiment, Sarah is happy to be your dance partner for a short while so you can see what happens. One of the other clinic rooms is free.'

Anna looked a little shy at first about this proposal, but then she said she would give it a go. They left the consulting room somewhat sheepishly but came back twenty minutes later after I had seen one of my follow-up cases.

'How did it go?' I was very curious to know.

'It was good . . . very good. Sarah is a great dancer and I didn't have any problem,' Anna said as she burst into sobs, while still smiling.

'Is that right, Sarah?' I enquired.

'She was perfect and there weren't any problems with her arm or leg.'

'Thank you so much for helping with this. It is a big deal for Anna.'

'It really is.' Anna wiped away her happy tears. 'You don't know how much.'

~ o ~

I saw Anna again a few more times over the years, but eventually it no longer became necessary for her to have regular assessments. A repeat MRI scan a year later showed no evidence of the cyst re-forming. She was able to go back to her normal life, including taking an active part in her dance club. She even entered competitions, but most of all she was extremely glad to be able to reconnect with her group of friends. Anna was back to her beautiful, happy self, her mother informed me, and that was all that she could ever wish for.

8

The self and identity

We started this book with a question: 'What makes us who we are?' Over the course of seven chapters, we've met people who became transformed by their neurological condition. David developed pathological apathy; Michael began to lose his semantic memory; Trish her episodic memory; Wahid suffered from visual hallucinations; Winston had impaired attention; Sue lacked control over her behaviour; and Anna lost the ability to know where her arm or leg were. How do these individuals help us to understand who we are? To answer this, it might first be worth considering the broader context in which others have tried to address the same question.

Philosophers have long considered the problem of *personal identity*, framing the discussion as one about *the self*. But despite centuries of debate, the self has turned out to be extremely elusive to pin down.[1] As Gordon Allport, an eminent psychologist put it:

Who is the I who knows the bodily me, who has an image of myself and sense of identity over time . . .? I know all these things and, what is more, I know that I know them. But who is

236

it who has this perspectival grasp . . .? It is much easier to feel the self than to define the self.[2]

In Western philosophy, René Descartes first developed the argument that the self cannot be a material, physical thing. In his view, the self is the mind, which is distinct from the brain or the body. What we sense, imagine and think – our mental acts – are experienced by the mind, not the brain. This 'Cartesian dualism' (the term for the mind–brain distinction, named after Descartes) raises some fundamental issues, not least of which is how the mind might interact with the brain. How can a non-physical entity such as the mind exert causal influence over a physical one? Or vice versa?

A very different position was adopted by John Locke, the English philosopher and physician. He was agnostic on whether the self had to be a non-physical form. In fact, he was willing to consider the possibility that it might be physical. For him, the key point was that the *personal identity* of an individuated, thinking person has to be mentally continuous over time. And for a person to be the same individual over time, there needs to be a continuity of consciousness. This, in turn, necessarily requires memory of past personal experiences.

Locke's position has often led to the view that memory is crucial to his view of personal identity. Without it, there couldn't be mental continuity. But it isn't clear that Locke meant that only the faculty of memory is crucial for personal identity. In fact, modern 'neo-Lockeans' have argued strongly that mental continuity requires more than just memory.[3] According to them, we need to invoke the entire breadth of psychological states that people are capable of when we consider what constitutes personal identity.

David Hume, the Scottish Enlightenment thinker, offered a different perspective. He considered the self simply to be an illusion. For Hume, the self is nothing more than 'a bundle' of experiences or perceptions. Personal identity is not a real entity but rather an illusion that we seem to be susceptible to.[4] Hume argued that it is our imagination that leads to the spurious appearance of a unitary persistence of mind or self. Memory is important, but only because without it there would be no semblance of causation in the perceptions we experience.[5] Several modern thinkers echo his view, but have taken the argument further.

The twentieth-century British philosopher Derek Parfit argued that psychological – and not any form of material or physical – continuity is what matters for personal survival,[6,7] while the American Daniel Dennett, like Hume, holds that the self is fiction. From his perspective, we create our selves through a self-narrative: 'We are all virtuoso novelists.' According to Dennett, 'We try to make all of our material cohere into a single good story. And that story is our autobiography. The chief fictional character at the center of that autobiography is one's self.' Dennett argues that we might consider the self in an analogous way to the 'center of gravity' of an object. It has no physical properties; it is not a real thing but it is a theorist's fiction. 'No one has ever seen or ever will see a center of gravity. As David Hume noted, no one has ever seen a self, either,' Dennett concludes.[8]

Not everyone would agree with him, of course. Some very eminent scholars, for example the philosopher Karl Popper and the Nobel Prize-winning neurophysiologist John Eccles, continued to defend a Cartesian dualist position in the modern era.[9] However, most contributors to this debate would nowadays subscribe to the view that the self and our sense of personal identity are emergent properties of the brain. Our mental selves are simply a creation of the activity of our brains.[10]

Notwithstanding this very reductionist perspective, social

psychologists remain convinced of the necessity to invoke the concept of a self. Roy Baumeister, for example, argues that we simply cannot ignore it. For him, while the self is not an object or a thing, it is a brain 'process' that is crucial for human existence.[11] According to Baumeister, 'selves exist in relation to a society'. The self is a cultural solution to the problem of how human societies succeed. Both individuals and the society they are embedded in are involved in shaping each self. Most humans, he and other social psychologists argue, have a *self-concept* – a mental representation of who they are.

This self-concept is a fundamental part of being a human. It is effectively a set of attributes that are characteristic of who we are. Crucially, it serves also to differentiate us from other people. Some researchers consider it as a kind of conceptual knowledge base that organises information about our selves, rather like we might organise knowledge about things in the external world.[12] When people are asked to write down statements that describe who they are, the results reveal that both their individual personality traits, as well their group memberships (e.g. 'I am a Catholic'), contribute to their concept of themselves.[13] Some psychologists also consider part of the self-concept to be a 'narrative self',[14] an internalised story about one's life that is used to understand who we are across time, not dissimilar in many ways to Dennett's view of the self.

This brief consideration of some Western concepts of the self reveals just how difficult it has been to have any consensus on what the self is. Other cultures, of course, have different views. Buddhists, for example, deny the existence of a self and consider personal identity to be delusional. Can modern neuroscience contribute to this debate? An intriguing question is whether it is possible to localise the self within the brain. With the advent of functional neuroimaging techniques, which allow us to measure activity in different brain regions while people perform

239

a cognitive task, this seemed a reasonable issue to tackle, at least for some neuroscientists. But how would we look for the self?

One way this has been attempted is to compare brain activity when people are asked to focus on their own emotional reactions to a set of images (self condition), then asked to make a different kind of judgement which doesn't refer to themself, such as whether the image shows an indoor or outdoor scene (non-self condition). The results of these types of brain-scanning investigations have led some researchers to conclude that there might be a set of cortical regions, in each cerebral hemisphere, located near the midline of the brain, which is consistently activated in self-referential conditions.[15]

Others, however, have critiqued the designs of these studies and pointed out that the activity that has been observed may not be attributed specifically to the 'self'.[16] Even if they were, they argue, the cortical regions that have been identified cover a very large swathe of the brain's midline structures (from front to back), and include other brain regions depending upon the particular task being performed. Most researchers have therefore concluded that there is no evidence that we can actually localise the self within the brain. This is the first major contribution of modern neuroscience to the long-standing debate about the self. Although, as Daniel Dennett pointed out years before: 'It is a category mistake to start looking around for the self in the brain.'[8] From his perspective, we're not going to find it because our self is made up by the entirety of our brain functions.

The second important contribution of neuroscience has been the understanding that focal brain lesions do not lead to complete loss of the self. Perhaps one of the most remarkable examples of this is the observations made on patients who have a rare surgical procedure performed on them to treat epilepsy. Some people whose seizures do not respond to medications have undergone the process of callosotomy (cutting of the corpus callosum, the huge bundle of fibres that connect the left and right cerebral hemispheres). The aim of this surgery is to stop the abnormal electrical activity of a seizure from crossing from one half of the brain to the other.

As a procedure to treat chronic, intractable epilepsy, such 'disconnection' of the left and right hemispheres from each other can be highly successful. Some of the individuals who have had this operation turned out also to be an invaluable set of people in whom to study functions of each hemisphere in semi-isolation. However, after surgery, one clear observation was that patients did not show any evidence in their everyday lives of a split self. They were not behaving in any way as if they were divided in two. As far as they were concerned, they were the same (unified) person.[6,10]

Similarly, people who have focal lesions of the brain, such as from a stroke or the excision of a tumour, can show a range of deficits depending upon the location of the brain damage, but there is no evidence that any of them lose their self as a result. They might become different selves as a consequence of their brain damage, but they do not lose an understanding of who they are. This is also the case in people who develop slowly progressive neurodegenerative conditions, such as Alzheimer's disease.[17] As we saw in Chapter 3, someone like Trish can start to lose their episodic memory (recollection of events) because of Alzheimer's, but they do not necessarily lose themselves. They know who they are. Although they may become a different individual, they retain a sense of their previous selves.

Such an observation is important in light of the influential proposal that Locke made in which memory is the glue that keeps a self together: he argued that it provides the ability to have a sense of a continuous personal identity. It has become clear, though, that most patients with amnesia (loss of episodic memory) still retain a sense of themselves, such as their personality traits.[18] As Stanley Klein and Cynthia Gangi put it: 'There are no documented cases in which a person has lost . . . self-knowledge while retaining other components of the self.'[19] Of course, in very advanced Alzheimer's disease, some patients may seem to be losing this ability, but by that stage it is more than their episodic memories that are impaired. The disease, which as we saw in Chapter Three, may often start in the entorhinal cortex adjacent to the hippocampus (**Fig. 5**). But eventually it will spread to affect all parts of the brain and therefore affect many different cognitive processes.

Even those people who we occasionally hear about in the news who are found, sometimes wandering on a beach, without any knowledge of who they are, do not really appear to have lost themselves for good. They have a diagnosis of what is called a fugue state, often triggered by very traumatic psychological experiences, and usually lasting only a few days or weeks. Exactly how personal identity and autobiographical memories become inaccessible or are effectively 'shut out' in such people is obviously of great interest. However, the temporary nature of their symptoms points to the fact that, even in these cases, the self is not permanently lost.

So it is also with the seven people we have met in this book. Each of their symptoms arose because of different cognitive processes becoming impaired. The conditions that led to the symptoms – whether they were a neurodegenerative condition such as Alzheimer's disease or a focal brain lesion such as a stroke or a brain cyst developing after traumatic brain injury – affected different brain systems. As a result, each individual became different, losing a piece of themselves.

What do these people's stories reveal about what makes us who we are? At one level, they demonstrate that the self – our personal identity – is composed of many different cognitive processes. If you lose one, you lose a particular faculty – such as episodic memory, or semantic memory, perception, attention, control over behaviour or representation of body parts. But crucially, they also show us that people don't lose their entire sense of self through the loss of any one of these cognitive modules.

In effect, these different cognitive processes together give rise to our self. To paraphrase Marvin Minsky, an American cognitive scientist and one of the pioneers of artificial intelligence, these cognitive processes are the fundamental entities from which minds are built: they are what constitute the 'Society of Mind'.[20] When part of that 'society' becomes dysfunctional, there is still a society left, but a different one: a different personal identity.

Identities are not just about individual personal identities, though. A crucial component of the self, some have argued, is social identity. Henri Tajfel and John Turner,[21] two very eminent social psychologists, proposed that *social identity* is how we define ourselves by our *relationships to others*, including the social groups we belong to.[22] Thus, whereas personal identity defines how a self ('I') is distinguished from other selves, social identity refers to how they are connected to other members of a group they belong to ('We'). All seven of the people we have encountered in this book changed in their personal identity, but this alteration also had a huge impact on their social identities.

David developed pathological apathy following his basal ganglia strokes. He lost his job and could not even be motivated to obtain social security benefits. His friends were kind enough to take him in to live with them. However, because he did not contribute to the domestic chores or engage with them, he began to become alienated. Thankfully, his levels of motivation were restored by a drug which stimulated dopamine receptors in the brain. He was able to return to his social network. David was lucky.

Michael, unfortunately, eventually became an outsider because of his lack of semantic knowledge. It had started with him losing the ability to appreciate humour, but slowly extended beyond this. It wasn't just complex concepts such as irony or embarrassment that he encountered difficulty in understanding. He began to forget the names of objects and crucially what they are and how they should be used. Slowly, his lack of conceptual knowledge pushed him from being an insider to becoming estranged from his own family.

Trish developed Alzheimer's disease, with loss of memory being the earliest manifestation. Her husband, Steve, found it increasingly difficult to cope with her amnesia, which included forgetting who he was – and even thinking he was a lover of hers. She would also confabulate, describing false recollections without being aware they were not true. She grew increasingly distant from her family and friends. Part of the problem was her denial that she had any difficulty with her memory. Once she acknowledged that she had Alzheimer's disease and was able to tell people, her relationships with them improved. They now had an explanation for her poor memory and adapted to this in their interactions with her.

Wahid, the bus driver who had originally come from Pakistan, developed visual hallucinations. He was worried that he was going mad and would therefore be interned in a psychiatric institution. When his friends and acquaintances found out, they made judgements about him and began to shun him. Fortunately, he eventually responded to a drug that increased the levels of the neurotransmitter acetylcholine in the brain, which allowed him to reconnect with people in his social network.

After his right parietal stroke, Winston, also a man who had been born in an ex-British colony (in his case, Jamaica), developed left-sided inattention. Originally of the Windrush generation, his Caribbean friends began to doubt that he had suffered a stroke because he didn't show the signs that they would have expected of such a condition. They even considered the possibility that he might have contracted syphilis but, once they were reassured that he hadn't, they made greater efforts to stay connected with him.

Sue developed frontotemporal dementia. Her behaviour changed remarkably. She became disinhibited, speaking her mind and blurting out comments in the street, wearing a cowgirl outfit to a hospital appointment, or summer clothes in the middle of winter. She lacked empathy and became impulsive in her habits, buying things she did not need. Her behaviour was callous, sometimes aggressive, and was considered offensive even by her close family members. A drug that affects the serotonin neuro-transmitter system was able to moderate some of her erratic behaviour and helped to restore better relationships, at least for a while.

Finally, Anna, a young woman who was born in Poland, started to lose awareness of her right arm and leg. This occurred as a result of an arachnoid cyst over her left parietal lobe. She had developed the cyst after a traumatic brain injury many years ago, following an unprovoked racist attack. Her symptoms began to be embarrassing for her as, without awareness and control over her hand, it would wander autonomously, touching people – including her dance partners – inappropriately. She became socially isolated. Fortunately, neurosurgical removal of the cyst was able to cure her of her symptoms and allow her to return to her fulfilling and lively social network.

Thus, all seven of the people we met underwent a change in their selves, which was associated with an alteration of both their personal and social identities. Some social psychologists argue that our self-concept (our beliefs about the personal qualities that make us who we are) has two distinct parts: personal identity and social identity. But according to the influential social psychologists Tajfel and Turner, it is not possible to separate our personal self from our social groups.[23] As David Carr, an American philosopher puts it: 'Personal identity is social identity.'[24]

Tajfel and Turner's research, along with that of others, has shown strikingly that simply dividing people who previously do not know each other arbitrarily into different groups leads to biases.[23] Individuals favour members of the same group (in-group favouritism), and act against those of another group (out-group prejudice). This occurs even when there is no motivation to do so, other than people's affiliation to a new group to which they have been randomly allocated in a research experiment. This process of *self-categorisation* appears to be an automatic, almost inbuilt, system that makes a person categorise themselves as belonging to a group and behave in a way that will advantage other members of that group, even if they are complete strangers. Group membership seems sufficient to generate identification with a collection of people and channel behaviour to its advantage.

According to social psychologists Baumeister and Leary, the need to belong to groups is fundamental to human existence.[25] It can lead to greater self-esteem and satisfaction with life, giving it more meaning and purpose. Some also argue that there is an evolutionary advantage for individuals to be embedded in socially integrated groups.[26] By being part of a network which cooperates for the greater good of the group, they are better protected, more likely to survive to adulthood, reproduce and raise offspring.

Without group memberships, many people become unhappy, distressed and sometimes unable to function. Their well-being is affected. Loneliness is associated with a significantly increased risk of coronary heart disease, stroke, depression, anxiety, greater cognitive decline and earlier death.[27,28] In some societies, when individuals do not conform to rules or cultural expectations, they are ostracised or shunned.[29] Externally imposed loneliness and social isolation is the punishment used to regulate individual behaviour.

The process of being able to identify socially with a group makes meaningful connections possible, allowing a person to benefit from fellow in-group members. For example, being part of a residents' association can allow people to feel safer. But how does a person gain group membership? Sometimes this may simply be due to circumstances. If you live in a particular area, and other residents are forming an association, you might be invited to join. If you start going to a university, you might be able to join the clubs associated with that organisation. If your parents are members of a religious group, you might be introduced to members of that community from childhood. However, it isn't always so easy.

Joining a group can be enormously challenging for a newcomer. First, they have to understand how the group works. What are the conventions and norms? How do people greet each other? Should one hug, kiss or shake hands? What are the attitudes and attributes of group members? What do they find funny? What are they serious about? What is not tolerated and what is accepted? Second, they need to establish good relations with group members, ideally influential ones who might lobby for their access to this network. Finally, they must show that they abide by the group 'culture' and its ethos.

None of this is straightforward. It requires the fundamental cognitive abilities that we have considered in this book: to perceive and attend successfully to important information that characterises a group; to understand its meaning and retain it in memory; and to have the motivation to use this knowledge in different social contexts without upsetting people.

The formidable obstacles to gaining entry into a social group are perhaps nowhere more apparent than in the case of immigrants to new countries. Worldwide, it is estimated that there

247

are over 280 million migrants (over 3.5 per cent of the global population).[30] They face enormous social barriers and, I would argue, significant cognitive challenges to integrate within an indigenous – or settled, established – community. Even if a person who has moved or been displaced to a new country wants to assimilate, there are many cultural differences which are not always easy to surmount. As Geert Hofstede, a Dutch social psychologist put it, culture is 'the collective programming of the mind that distinguishes the members of one category of people from another'.[31]

A newcomer needs to have the cognitive abilities to repro-gramme their mind so that it is different from the one they have grown up with if they are to assimilate or integrate successfully. Some people are capable of doing this, successfully adapting to new circumstances to allow them to gain insider status – to 'fit' into a group. But not everyone is. Even if they are, the route to becoming a member of a new community is not always made easy. Stereotyping and stigmatising of immigrants by those who see themselves as 'locals' can make assimilation extremely difficult.

Three of the people we met in this book – Wahid, Winston and Anna – were confronted by what many generations of *Bloody Foreigners*[32] had to contend with when they first came to Britain. Resentment, prejudice and even violence is common when people from different cultures meet. Winston was involved in the 1950s race riots when gangs of white youths descended on Black people in Notting Hill. Wahid had been assaulted when working as a road sweeper, accused of robbing people of their jobs. And Anna had been severely attacked simply because she was speaking on her mobile phone in Polish to a friend.

Why does this happen? How do we explain prejudice and bias when individuals don't even know the people they attack? One influential explanation that has emerged is from the social identity theory proposed by Tajfel and Turner.[21] They argue that, even without knowing someone, people evaluate others at the level of the groups they belong to. A person's social identity, according to Tajfel and Turner, depends on determining how similar or different they are from someone else. Without 'others' who are different from us, we can't claim to be part of a group with a particular identity. In an attempt to gain a positive social identity, improve self-esteem and sense of group worth, individuals seek to differentiate their in-group positively from out-groups, which they cast in a negative light.

Prejudice and discrimination targeted at out-groups arise from conflicts of interest between one or more groups when they appear to be in competition for key resources, such as jobs and wealth, status and prestige. Because a person's sense of self is inextricably tied into their group identity, people judge individuals by the groups they belong to, without knowing anything about their personal attributes and qualities. This type of group-level comparison, it has been argued, is at the heart of conflict between people of different backgrounds.

For newcomers attempting to gain entry into an established group, therefore, the evaluation they undergo is not simply about themselves but perhaps more about the group they belong to, and the strength of their affiliation to it. If they do successfully gain access to a group, remaining an insider isn't guaranteed. Crucially, retaining membership, even if this has been long-standing, requires people to continue to conform to the norms: the rules or standards of behaviour that are accepted by the group. When they cease to do so, their relationship with other people comes under threat. They risk their existence within a group.

The seven people we met in this book were all confronted by this very real possibility because their behaviour had changed so significantly. As a consequence of the cognitive effects of their brain disorder, they were no longer considered acceptable within their social networks. An important corollary to this is that the cognitive processes that became dysfunctional in these individuals are normally crucial to maintaining our social identity – our relationships to other people – as well as our personal identity.

Fundamental cognitive functions such as perception, attention, episodic and semantic memory, motivation, control over behaviour and the body schema all contribute to our identities. Personality traits and emotional responses are, of course, also important in defining the self. But the seven remarkable individuals we have met in this book poignantly reveal how even very basic brain functions play a key role in determining who we are. They are crucial parts of the 'society of the mind' that creates our self, but they are also crucial to keeping us within society.

Acknowledgements

This book would not have been possible without the patients I have met in my clinics over the last thirty years. To them, their families and friends, I extend my heartfelt thanks for sharing their stories, and allowing me to be involved in their care. To maintain confidentiality, I have altered the personal backgrounds and circumstances, as well as the timelines of clinical assessments, of the people described in this book.

My thanks also to the many colleagues in the clinical and academic institutions where I learned about neuroscience, psychology and how to be a clinician. They include my tutors at Oxford University who inspired me to think adventurously; my colleagues from the Massachusetts Institute of Technology who showed me how things can get done if you put your mind to it; and neurologists and neuroscientists at Imperial College who gave me the opportunity to do clinical science in a way that I had not experienced before. I am particularly grateful to my colleagues and friends from Queen Square, London, where this book is set. They include the clinicians, neuroscientists, psychologists and administrators at the Department of Brain Repair & Rehabilitation in the UCL Institute of Neurology, and from the UCL Institute of Cognitive Neuroscience. Together with my colleagues at Oxford – neurologists, psychologists, psychiatrists, administrators, nurses and secretaries – they have been instrumental in my progress and in giving me the motivation to write this book.

None of my own research (some of which is described here) would have been possible without the dedicated work of the students, postdocs and research coordinators who have been involved in my team. Many of them sparked innovations and new ways of thinking that have had a deep influence on me. I'm grateful to you all, and thank you for putting up with some of my madcap ideas. Thanks also to the Wellcome Trust for supporting my research for over twenty-five years, and for giving me the – rare – opportunity to work as a clinician and a scientist on projects that I am passionate about.

I am grateful also to Simon Thorogood, my editor, and the team at Canongate; my copy-editor, Emma Hargrave; Jessica Woollard, my agent; Yehrin Tong, my cover illustrator; Frank Tallis who was my writing mentor; Bobby Nayyar at Spread the Word; Francesca Barrie and Ellen Johl at The Wellcome Collection; and my fellow awardees on the Wellcome Collection / Spread the Word non-fiction programme. Thank you for all your thoughtful input on this project, and for helping me to navigate the publishing world.

Finally, my thanks to my parents and family for giving me the possibilities that I have been lucky enough to experience. I am particularly grateful to Claire, Megan and Adam for their support, careful reading and feedback on early chapter drafts. None of this, along with many other achievements, would have been possible without your encouragement and forbearance.

Further reading

For more on social psychology and philosophical thinking on the self

Baumeister, R. F. *The Self Explained. Why and How We Become Who We Are.* New York: The Guilford Press; 2022.

Birney, M. E. *Self and Identity. The Basics.* Oxford: Routledge; 2023.

Dainton, B. *Self: What Am I?* Milton Keynes: Penguin; 2014.

Hewstone, M., and Stroebe, W., eds. *An Introduction to Social Psychology.* 7th ed. Hoboken, NJ: Wiley; 2020.

Jenkins, R. *Social Identity.* Abingdon: Routledge; 2014.

Smith, E. R., Mackie, D. M., and Claypool, H. M. *Social Psychology.* 4th ed. Abingdon: Routledge; 2019.

For more on the neuroscience background

Kandel, E., Koester, J. D., Mack, S. H., and Siegelbaum, S. A., eds. *Principles of Neural Science.* 6th ed. New York: McGraw Hill; 2021.

Purves, D., LaBar, K. S., Woldroff, M., Cabeza, R., and Huettel, S. A. *Principles of Cognitive Neuroscience.* 2nd ed. Oxford: Sinauer; 2013.

For more on cognitive neurology and dementia

Husain, M., and Schott J.M., eds. *Oxford Textbook of Cognitive Neurology and Dementia* Oxford: Oxford University Press; 2016.

References

Introduction

1. Kear, J. 'Une Chambre Mentale: Proust's Solitude.' In: *Writers' Houses and the Making of Memory*. New York: Routledge; 2007, pp. 221–235.
2. Husain, M. 'Proust and his neurologists: the challenge of functional disorders.' *Brain*. 2021; 144(8): 2227.
3. Tadié, J-Y. *Marcel Proust. A Life*. New York: Penguin Books; 2000.
4. Painter, G. D. *Marcel Proust. A Biography. Vol 2*. London: Chatto & Windus; 1967.
5. Husain, M., and Schott J. M., eds. *Oxford Textbook of Cognitive Neurology and Dementia*. Oxford: Oxford University Press; 2016.
6. Shattuck, R. *Proust's Way: A Field Guide to In Search of Lost Time*. New York: W. W. Norton & Co; 2000.
7. Baumeister, R. F., and Leary, M. R. 'The need to belong: desire for interpersonal attachments as a fundamental human motivation.' *Psychological Bulletin*. 1995; 117(3): pp. 57–89.
8. Allen, K. A. *The Psychology of Belonging*. London: Taylor and Francis; 2020.
9. Kuhn, M. H., and McPartland T. S. 'An Empirical Investigation of Self-Attitudes.' *American Sociology Review*. 1954; 19(1): pp. 68–76.
10. Zurcher, L. *The Mutable Self: A Self-Concept for Social Change*. Beverly Hills, CA: SAGE Publications; 1977.

11. Tajfel, H. *Social Categorization, Social Identity and Social Comparison*. London: Academic Press; 1978.

12. Ansell, N. *Deep Country: Five Years in the Welsh Hills*. London: Penguin Books; 2012.

13. Williams, K. D., and Zadro, L. 'Ostracism: On Being Ignored, Excluded, and Rejected.' In: Leary, M. R., ed. *Interpersonal Rejection*. Oxford: Oxford University Press; 2012, pp. 21–54.

14. Baumeister, R. F. *The Self Explained. Why and How We Become Who We Are*. New York: The Guilford Press; 2022.

1 | A little miracle

1. Shorvon, S., and Compston, A. *Queen Square: A History of the National Hospital and Its Institute of Neurology*. Cambridge: Cambridge University Press, 2018.

2. Macalpine, I., and Hunter, R. 'The "Insanity" of King George III: A Classic Case of Porphyria.' *British Medical Journal*. 1966; 1(5479): pp. 65–71.

3. Peters, T. J., and Beveridge, A. 'The madness of King George III: A psychiatric re-assessment.' *History of Psychiatry*. 2010; 21(1): pp. 20–37.

4. Arnold, C. *Bedlam. London and Its Mad*. London: Simon & Schuster; 2008.

5. Wilson, S. A. 'The Croonian Lectures: On some disorders of motility and of muscle tone: with special reference to the corpus striatum.' *Lancet*. 1925; 206(5314): pp. 1–10.

6. Alexander, G. E, DeLong, M. R., and Strick, P. L. 'Parallel organization of functionally segregated circuits linking basal ganglia and cortex.' *Annual Review of Neuroscience*. 1986; 9: pp. 357–381.

7. Haber, S. N., and Knutson, B. 'The reward circuit: linking primate anatomy and human imaging.' *Neuropsychopharmacology*. 2010; 35(1): pp. 4–26.

8. Mink, J. W. 'The basal ganglia.' In: Squire, L. R., Berg, D., Bloom, F. E., Du Lac, S., Ghosh, A., and Spitzer, N. C., eds. *Fundamental Neuroscience*. 4th ed. Waltham, MA: Academic Press; 2012: pp. 653–676.

9. Panigrahi, B., Martin, K. A., Li. Y., et al. 'Dopamine Is Required for the Neural Representation and Control of Movement Vigor.' *Cell.* 2015; 162(6): pp. 1418–1430.

10. Dudman, J. T., Krakauer, J. W. 'The basal ganglia: from motor commands to the control of vigor.' *Current Opinion in Neurobiology.* 2016; 37: pp. 158–166.

11. Klaus, A., Alves Da Silva, J., and Costa, R. M. 'What, If, and When to Move: Basal Ganglia Circuits and Self-Paced Action Initiation.' *Annual Review of Neuroscience.* 2019; 42: pp. 459–483.

12. Mogenson, G. J., Jones, D. L., and Yim, C. Y. 'From motivation to action: functional interface between the limbic system and the motor system.' *Progress in Neurobiology.* 1980; 14(2-3): pp. 69–97.

13. Salamone, J. D., Yohn, S. E., López-Cruz, L., San Miguel, N., and Correa, M. 'Activational and effort-related aspects of motivation: neural mechanisms and implications for psychopathology.' *Brain.* 2016; 139(5): pp. 1325–1347.

14. Le Heron, C., Holroyd, C. B., Salamone, J., and Husain, M. 'Brain mechanisms underlying apathy.' *Journal of Neurology, Neurosurgery and Psychiatry.* 2019; 90(3).

15. Dunbar, R., Barrett, L., and Lycett, J. *Evolutionary Psychology. A Beginner's Guide.* London: Oneworld Publications; 2007.

16. Adam, R., Leff, A., Sinha, N., et al. 'Dopamine reverses reward insensitivity in apathy following globus pallidus lesions.' *Cortex.* 2013; 49(5): pp. 1292–1303.

17. Salamone, J. D., and Correa, M. 'The mysterious motivational functions of mesolimbic dopamine.' *Neuron.* 2012; 76(3): pp. 470–485.

18. Sacks, O. *Awakenings.* London: Duckworth; 1973.

19. Husain, M, and Roiser, J. P. 'Neuroscience of apathy and anhedonia: a transdiagnostic approach.' *Nature Reviews Neuroscience.* 2018; 19(8): pp. 470–484.

20. Saleh, Y., Le Heron, C., Petitet, P, et al. 'Apathy in small vessel cerebrovascular disease is associated with deficits in effort-based decision making.' *Brain.* 2021; 144(4): pp. 1247–1262.

21. Bonnelle, V., Manohar, S., Behrens, T., and Husain, M. 'Individual Differences in Premotor Brain Systems Underlie Behavioral Apathy.' *Cerebral Cortex*. 2016; 26(2): pp. 807–819.

22. Costello, H., Husain, M., and Roiser, J. P. 'Apathy and Motivation: Biological Basis and Drug Treatment.' *Annual Review of Pharmacology and Toxicology*. 2023; 64: pp. 313–338.

2 | The man who ran out of words

1. Lynch, J., ed. *Samuel Johnson's Dictionary: Selections from the 1755 Work That Defined the English Language*. 2nd ed. London: Atlantic Books; 2004.

2. Damrosch, L. *The Club*. New Haven: Yale University Press; 2019.

3. Hibbert, C. *The Personal History of Samuel Johnson*. 2nd ed. New York: Harper & Row; 1971.

4. Pearce, J. M. S. 'Doctor Samuel Johnson: "the Great Convulsionary" a victim of Gilles de la Tourette's syndrome.' *Journal of the Royal Society of Medicine*. 1994; 87(7): pp. 396–399.

5. Collins, A. M., and Quillian, M. R. 'Retrieval time from semantic memory.' *Journal of Verbal Learning and Verbal Behavior*. 1969; 8(2): pp. 240–247.

6. McClelland, J. L., and Rogers, T. T. 'The parallel distributed processing approach to semantic cognition.' *Nature Reviews Neuroscience*. 2003; 4(4): pp. 310–322.

7. Tulving, E. 'Episodic and semantic memory.' In: *Organization of Memory*. New York: Academic Press; 1972: pp. 381–403.

8. Rascovsky, K., Growdon, M. E., Pardo, I. R., Grossman, S., and Miller, B. L. '"The quicksand of forgetfulness": Semantic dementia in *One Hundred Years of Solitude*.' *Brain*. 2009; 132(9): pp. 2609–2616.

9. Hodges, J. R., Patterson, K., Oxbury, S., and Funnell, E. 'Semantic dementia: Progressive fluent aphasia with temporal lobe atrophy.' *Brain*. 1992; 115(6): pp. 1783–1806.

10. Hodges, J. R., and Patterson, K. 'Semantic dementia: a unique clinicopathological syndrome.' *Lancet Neurology*. 2007; 6(11): pp. 1004–1014.

11. Lambon Ralph, M. A., Jefferies, E., Patterson, K., and Rogers, T. T. 'The neural and computational bases of semantic cognition.' *Nature Reviews Neuroscience*. 2016; 18(1): pp. 42–55.

12. Warrington, E. K., and Shallice, T. 'Category specific semantic impairments.' *Brain*. 1984; 107(3): pp. 829–853.

13. Kemmerer, D. *Cognitive Neuroscience of Language*. 2nd ed. New York: Routledge; 2022.

14. Patterson, K., Nestor, P. J., and Rogers, T. T. 'Where do you know what you know? The representation of semantic knowledge in the human brain.' *Nature Review Neuroscience*. 2007; 8(12): pp. 976–987.

15. Kroeger, P. *Analyzing Meaning: An Introduction to Semantics and Pragmatics*. 3rd ed. Berlin: Language Science Press; 2018.

16. Raskin, V. *Semantic Mechanisms of Humor*. Springer Netherlands; 1984.

17. Dunbar, R. I. M. 'Laughter and its role in the evolution of human social bonding.' *Philosophical Transactions of the Royal Society of London Biological Sciences*. 2022; 377(1863).

18. Rohrer, J. D, Lashley, T., Schott, J. M., et al. 'Clinical and neuroanatomical signatures of tissue pathology in frontotemporal lobar degeneration.' *Brain*. 2011; 134(9): pp. 2565–2581.

3 | Losing my memory?

1. Carson, A. 'Capgras syndrome.' *Brain*. 2023; 146(10):3955–3957.

2. Allen, T. A., Fortin, N. J. 'The evolution of episodic memory.' *Proceedings of the National Academy of Sciences U S A*. 2013; 110(SUPPL2): pp. 10379–10386.

3. Corkin, S. *Permanent Present Tense: The Man with No Memory, and What He Taught the World*. London: Penguin; 2014.

4. Josselyn, S. A., and Tonegawa, S. 'Memory engrams: Recalling the past and imagining the future.' *Science*. 2020; 367(6473): eaaw4325.

5. Kandel, E. R. *In Search of Memory: The Emergence of a New Science of Mind*. New York: W. W. Norton & Co; 2007.

6. Kandel, E. R. Nobel Lecture 2000. https://www.nobelprize. org/prizes/medicine/2000/kandel/lecture/

7. Squire, L. R., and Zola, S. M. 'Structure and function of declarative and nondeclarative memory systems.' *Proceedings of the National Academy of Sciences U S A*. 1996; 93(24): pp. 13515–13522.

8. Bartlett, F. C. *Remembering*. Cambridge: Cambridge University Press; 1932.

9. Brewer, W. F., and Treyens, J. C. 'Role of schemata in memory for places.' *Cognitive Psychology*. 1981; 13(2): pp. 207–230.

10. Loftus, E. F., and Palmer, J. C. 'Reconstruction of automobile destruction: An example of the interaction between language and memory.' *Journal of Verbal Learning and Verbal Behavior*. 1974; 13(5): pp. 585–589.

11. Wells, G. L., and Bradfield, A. L. '"Good, you identified the suspect": Feedback to eyewitnesses distorts their reports of the witnessing experience.' *Journal of Applied Psychology*. 1998; 83(3): pp. 360–376.

12. Schnider, A. *The Confabulating Mind*. Oxford: Oxford University Press; 2013.

13. Tabi, Y. A., Husain, M. 'Clinical assessment of parietal lobe function.' *Practical Neurology*. 2023; 23(5): pp. 404–407.

14. Bruner, E., Battaglia-Mayer, A., and Caminiti, R. 'The parietal lobe evolution and the emergence of material culture in the human genus.' *Brain Structure and Function*. 2022; 228: pp. 145–167.

15. Ryan, N. S., Rossor, M. N., and Fox, N. C. 'Alzheimer's disease in the 100 years since Alzheimer's death.' *Brain*. 2015; 138(12): pp. 3816–3821.

16. Hardy, J. A., and Higgins, G. A. 'Alzheimer's disease: The amyloid cascade hypothesis. *Science*. 1992; 256(5054): pp. 184–185.

17. Roy, D. S., Arons, A., Mitchell, T. I., Pignatelli, M., Ryan, T. J.,

and Tonegawa, S. 'Memory retrieval by activating engram cells in mouse models of early Alzheimer's disease.' *Nature*. 2016; 531(7595): pp. 508–512.

18. Ramirez, S., Liu, X., Lin, P. A., et al. 'Creating a false memory in the hippocampus.' *Science*. 2013; 341(6144): pp. 387–391.

19. Delbourgo, J. *Collecting the World: The Life and Curiosity of Hans Sloane*. London: Penguin; 2017.

4 | Visitors in the night

1. Porter, R. *Madness: A Brief History*. Oxford: Oxford University Press; 2002.

2. Harrison, P., Cowen, P., Burns, T., and Fazel, M. *Shorter Oxford Textbook of Psychiatry*. 7th ed. Oxford: Oxford University Press; 2017.

3. Khalifa, N., Hardie, T. 'Possession and Jinn.' *Journal of the Royal Society of Medicine*. 2005; 98(8): pp. 351–353.

4. Khalifa, N., Hardie, T., and Mullick, M. S. I. *Jinn and Psychiatry: Comparison of Beliefs among Muslims in Dhaka and Leicester*. London: Royal College of Psychiatrists; 2012.

5. Sikander, S. 'Pakistan.' *Lancet Psychiatry*. 2020; 7(10): p. 845.

6. Cooper, C., Spiers, N., Livingston, G., et al. 'Ethnic inequalities in the use of health services for common mental disorders in England.' *Social Psychiatry and Psychiatric Epidemiology*. 2013; 48(5): pp. 685–692.

7. Chan, D., Rossor, M. N. '"—but who is that on the other side of you?" Extracampine hallucinations revisited.' *Lancet*. 2002; 360(9350): pp. 2064–2066.

8. Boring, E. G. *A History of Experimental Psychology*. Engelwood Cliffs, New Jersey: Pretnice-Hall; 1957.

9. Dayan, P., Hinton, G. E., Neal, R. M., and Zemel R. S. 'The Helmholtz machine.' *Neural Computation*. 1995; 7(5): pp. 889–904.

10. Corlett, P. R., Horga, G., Fletcher, P. C., Alderson-Day, B., Schmack, K., and Powers, A. R. 'Hallucinations and Strong Priors.' *Trends in Cognitive Sciences*. 2019; 23(2): pp. 114–127.

11. Engelhardt, E., and Gomes, M da M. 'Lewy and his inclusion bodies: Discovery and rejection.' *Dementia & Neuropsychology*. 2017; 11(2): pp. 198–201.

12. Koga, S., Sekiya, H., Kondru, N., Ross, O. A., and Dickson, D. W. 'Neuropathology and molecular diagnosis of Synucleinopathies.' *Molecular Neurodegeneration*. 2021; 16(1).

13. McKeith, I. G. 'Dementia with Lewy bodies and Parkinson's disease with dementia: Where two worlds collide.' *Practical Neurology*. 2007; 7(6): pp. 374–382.

14. Okkels, N., Horsager, J., Labrador-Espinosa, M., et al. 'Severe cholinergic terminal loss in newly diagnosed dementia with Lewy bodies.' *Brain*. 2023; 146(9): pp. 3690–3704.

15. Zarkali, A., Adams, R. A., Psarras, S., Leyland, L. A., Rees, G., and Weil, R. S. 'Increased weighting on prior knowledge in Lewy body–associated visual hallucinations.' *Brain Communications*. 2019; 1(1).

16. Mehraram, R., Peraza, L. R., Murphy, N. R. E., et al. 'Functional and structural brain network correlates of visual hallucinations in Lewy body dementia.' *Brain*. 2022; 145(6): pp. 2190–2205.

17. Porter, R. *Madmen: A Social History of Madhouses, Mad-Doctors & Lunatics*. Stroud: Tempus; 2006.

18. Andrews, J., and Scull, A. *Undertaker of the Mind. John Monro and Mad-Doctoring in Eighteenth-Century England*. Berkeley & Los Angeles: University of California Press; 2001.

19. Arnold, C. *Bedlam. London and Its Mad*. London: Simon & Schuster; 2008.

20. Ramscar, M. *George III's Illnesses and His Doctors*. Barnsley, Yorks: Pen & Sword Books; 2023.

21. Higgins, E., and George, M. S. *The Neuroscience of Clinical Psychiatry: The Pathophysiology of Behavior and Mental Illness*. 3rd ed. Philadelphia: Lippincott Williams and Wilkins; 2018.

22. Foucault, M. *Madness and Civilization: A History of Insanity in the Age of Reason*. New York: Pantheon Books; 1965.

23. Berry, J. W., Poortinga, Y. H., Breugelmans, S. M., Chasiotis, A.,

and Sam, D. L. *Cross-Cultural Psychology: Research and Applications.* 3rd ed. Cambridge: Cambridge University Press; 2011.

24. Burn, D., Emre, M., McKeith, I., et al. 'Effects of rivastigmine in patients with and without visual hallucinations in dementia associated with Parkinson's disease.' *Movement Disorders.* 2006; 21(11): pp. 1899–1907.

25. Gratwicke, J., Zrinzo, L., Kahan, J., et al. 'Bilateral Deep Brain Stimulation of the Nucleus Basalis of Meynert for Parkinson Disease Dementia: A Randomized Clinical Trial.' *JAMA Neurology.* 2018; 75(2): pp. 169–178.

26. Gratwicke, J., Zrinzo, L., Kahan, J., et al. 'Bilateral nucleus basalis of Meynert deep brain stimulation for dementia with Lewy bodies: A randomised clinical trial.' *Brain Stimulation.* 2020; 13(4): pp. 1031–1039.

5 | Neglecting me quietly

1. Brain, W. R. 'Visual disorientation with special reference to lesions of the right cerebral hemisphere.' *Brain.* 1941; 64(4): pp. 244–272.

2. Mcfie, J., Piercy, M. F., and Zangwill, O. L. 'Visual-spatial agnosia associated with lesions of the right cerebral hemisphere.' *Brain.* 1950; 73(2): pp. 167–190.

3. Hillis, A. E. 'Neurobiology of unilateral spatial neglect.' *Neuroscientist.* 2006; 12(2): pp. 153–163.

4. Esposito, E., Shekhtman, G., and Chen, P. 'Prevalence of spatial neglect post-stroke: A systematic review.' *Annals of Physical Rehabilitation Medicine.* 2021; 64(5): 101459.

5. Desimone, R., and Duncan, J. 'Neural mechanisms of selective visual attention.' *Annual Review of Neuroscience.* 1995; 18: pp. 193–222.

6. Corbetta, M., and Shulman, G. L. 'Control of goal-directed and stimulus–driven attention in the brain.' *Nature Reviews Neuroscience.* 2002; 3(3): pp. 201–215.

7. Posner, M. I. 'Orienting of attention.' *Quarterly Journal of Experimental Psychology.* 1980; 32(1): pp. 3–25.

8. Husain, M., and Rorden, C. 'Non-spatially lateralized mechanisms in hemispatial neglect.' *Nature Reviews Neuroscience.* 2003; 4(1).

9. Parton, A., Malhotra, P., and Husain, M. 'Hemispatial neglect.' *Journal of Neurology, Neurosurgery and Psychiatry.* 2004; 75(1): pp. 13–21.

10. Cantagallo, A, and Sala, S. D. 'Preserved insight in an artist with extrapersonal spatial neglect.' *Cortex.* 1998; 34(2): pp. 163–189.

11. Husain, M., and Stein, J. 'Rezsö Bálint and His Most Celebrated Case.' *Archives of Neurology.* 1988; 45(1).

12. Glickstein, M., and Whitteridge, D. 'Tatsuji Inouye and the mapping of the visual fields on the human cerebral cortex.' *Trends in Neurosciences.* 1987; 10: pp. 350–353.

13. Alvis-Miranda, H. R., Rubiano, A. M., Agrawal, A., et al. 'Craniocerebral Gunshot Injuries; A Review of the Current Literature.' *Bulletin of Emergency and Trauma.* 2016; 4(2): pp. 65–74.

14. Holmes, G., and Lister, W. 'Disturbances of vision from cerebral lesions, with special reference to the representations of the macula.' *Brain.* 1916; 39(1–2): pp. 34–73.

15. Holmes, G. 'Disturbances of visual orientation.' *British Journal of Ophthalmology.* 1918; 2(9): p. 449.

16. Holmes, G., and Horrax, G. 'Disturbances of spatial orientation and visual attention, with loss of stereoscopic vision.' *Archives of Neurology and Psychiatry.* 1919; 1: pp. 385–407.

17. Trojano, L. 'Constructional apraxia from the roots up: Kleist, Strauss, and their contemporaries.' *Neurological Sciences.* 2020; 41(4): pp. 981–988.

18. Hecaen, H., Penfield, W., Bertrand, C., and Malmo, R. 'The syndrome of apractognosia due to lesions of the minor cerebral hemisphere.' *AMA Archives of Neurology and Psychiatry.* 1956; 75(4): pp. 400–434.

19. De Renzi, E. *Disorders of Space Exploration and Cognition.* Chichester: John Wiley & Sons, Ltd; 1982.

20. Dalmaijer, E. S., Li, K. M. S., Gorgoraptis, N., et al.

'Randomised, double-blind, placebo-controlled crossover study of single-dose guanfacine in unilateral neglect following stroke.' *Journal of Neurology, Neurosurgery and Psychiatry*. 2018; 89(6): pp. 593–598.

21. Gorgoraptis, N., Mah, Y-H., MacHner, B., et al. 'The effects of the dopamine agonist rotigotine on hemispatial neglect following stroke.' *Brain*. 2012; 135(8): pp. 2478–2491.

22. Swayne, O. B., Gorgoraptis, N., Leff, A., and Ajina, S. 'Exploring the use of dopaminergic medication to treat hemispatial inattention during in–patient post-stroke neurorehabilitation.' *Journal of Neuropsychology*. 2022; 16(3): pp. 518–536.

23. Davis, G. *'The Cruel Madness of Love': Sex, Syphilis and Psychiatry in Scotland, 1880–1930*. Amsterdam: Editions Rodopi B.V.; 2008.

24. Rees, G., Wojciulik, E., Clarke, K., Husain, M., Frith, C., and Driver, J. 'Unconscious activation of visual cortex in the damaged right hemisphere of a parietal patient with extinction.' *Brain*. 2000; 123(8).

25. Marshall, J. C., Halligan, P. W. 'Blindsight and insight in visuospatial neglect.' *Nature*. 1988; 336(6201): pp. 766–767.

6 | The woman who didn't care

1. Macmillan, M. *An Odd Kind of Fame: Stories of Phineas Gage*. Cambridge, Mass.: MIT Press; 2002.

2. Eslinger, P. J., and Damasio, A. R. 'Severe disturbance of higher cognition after bilateral frontal lobe ablation: patient EVR.' *Neurology*. 1985; 35(12): pp. 1731–1741.

3. Teuber, H. L. 'The riddle of frontal lobe function in man.' *Neuropsychology Review*. 2009; 19(1): pp. 25–46.

4. Badre, D. *On Task. How Our Brain Gets Things Done*. Princeton: Princeton University Press; 2020.

5. Lezak, M., Howieson, D., Bigler, E., and Tranel, D. *Neuropsychological Assessment* 5th ed. New York: Oxford University Press; 2012.

6. Gilbert, S. J., and Burgess, P. W. 'Executive function.' *Current Biology*. 2008; 18(3): pp. R110–R114.

7. Koechlin, E., and Summerfield, C. 'An information theoretical approach to prefrontal executive function.' *Trends in Cognitive Sciences*. 2007; 11(6): pp. 229–235.

8. Badre, D., and Desrochers, T. M. 'Hierarchical cognitive control and the frontal lobes.' *Handbook of Clinical Neurology*. 2019; 163: pp. 165–177.

9. Damasio, A. R. 'The somatic marker hypothesis and the possible functions of the prefrontal cortex.' *Philosophical Transactions of the Royal Society of London Biological Sciences*. 1996; 351(1346): pp. 1413–1420.

10. Koenigs, M., Young, L., Adolphs, R., et al. 'Damage to the prefrontal cortex increases utilitarian moral judgements.' *Nature*. 2007; 446(7138): pp. 908–911.

11. Anderson, S. W., Bechara, A., Damasio, H., Tranel, D., and Damasio, A. R. 'Impairment of social and moral behavior related to early damage in human prefrontal cortex.' *Nature Neuroscience*. 1999; 2(11): pp. 1032–1037.

12. Stuss, D. T., Gallup, G. G., and Alexander, M. P. 'The frontal lobes are necessary for "theory of mind."' *Brain*. 2001; 124(Pt 2): pp. 279–286.

13. Mendez, M. F., and Shapira, J. S. 'Altered emotional morality in frontotemporal dementia.' *Cognitive Neuropsychiatry*. 2009; 14(3): pp. 165–179.

14. Zago, S., Scarpazza, C., Difonzo, T., et al. 'Behavioral Variant of Frontotemporal Dementia and Homicide in a Historical Case.' *Journal of the American Academy of Psychiatry and the Law*. 2021; 49(2): pp. 219–227.

15. Rankin, K. P., Gorno-Tempini, M. L., Allison, S. C., et al. 'Structural anatomy of empathy in neurodegenerative disease.' *Brain*. 2006; 129(11): pp. 2945–2956.

16. Lebert, F., Stekke, W., Hasenbroekx, C., and Pasquier, F. 'Frontotemporal dementia: a randomised, controlled trial with trazodone.' *Dementia and Geriatric Cognitive Disorders*. 2004; 17(4): pp. 355–359.

17. Hughes, D., and Mallucci, G. R. 'The unfolded protein response in neurodegenerative disorders – therapeutic modulation of the PERK pathway.' *FEBS Journal.* 2019; 286(2): pp. 342–355.

7 | Is that your hand or mine?

1. Penfield, W., and Boldrey, E. 'Somatic motor and sensory representation in the cerebral cortex of man as studied by electrical stimulation.' *Brain.* 1937; 60(4): pp. 389–443.

2. Denny Brown, D., and Banker, B. Q. 'Amorphosynthesis from left parietal lesion.' *AMA Archives of Neurology and Psychiatry.* 1954; 71(3): pp. 302–313.

3. Lillie, M. C. 'Cranial surgery dates back to Mesolithic.' *Nature.* 1998; 391(6670): p. 854.

4. Head, H., and Holmes, G. 'Sensory disturbance from cerebral lesions.' *Brain.* 1911; 34(2–3): pp. 102–254.

5. Wolpert, D. M., Goodbody, S. J., and Husain, M. 'Maintaining internal representations: The role of the human superior parietal lobe.' *Nature Neuroscience.* 1998; 1(6).

6. Costandi, M. *Body Am I.* Cambridge, Mass.: MIT Press; 2022.

7. Cole, J. *Losing Touch: A Man without His Body.* Oxford: Oxford University Press; 2016.

8. Gallagher, S., and Cole, J. 'Body schema and body image in a deafferented subject.' *Journal of Mind and Behavior.* 1995; 16: pp. 369–390.

9. Cassam, Q. 'The Embodied Self.' In: *The Oxford Handbook of the Self.* Oxford: Oxford University Press; 2011: pp. 139–156.

8 | The self and identity

1. Dainton, B. *Self: What Am I?* Milton Keynes: Penguin; 2014.

2. Allport, G. W. *Patterns and Growth in Personality.* New York: Holt, Rinehart & Winston; 1961.

3. Shoemaker, S. 'On What We Are.' In: Gallagher, S., ed. *The Oxford Handbook of the Self.* Oxford: Oxford University Press; 2011: pp. 352–371.

4. Hood, B. *The Self Illusion*. London: Constable; 2011.

5. Thiel, U. 'Hume and the belief in personal identity.' In: *The Early Modern Subject: Self-Consciousness and Personal Identity from Descartes to Hume*. Oxford: Oxford University Press; 2011: pp. 382–406.

6. Parfit, D. 'Personal Identity.' *Phisosophical Reviews*. 1971; 80(1): pp. 3–27.

7. Parfit, D. *Reasons and Persons*. Oxford: Oxford University Press; 1986.

8. Dennett, D. 'The Self as a Center of Narrative Gravity.' In: Kessel, F., Cole, P., and Johnson, D., eds. *Self and Consciousness: Multiple Perspectives*. Hillsdale, NJ: Lawrence Erlbaum Associates; 1992: pp. 103–115.

9. Popper, K. R., and Eccles, J. C. *The Self and Its Brain: An Argument for Interactionism*. Berlin: Springer-Verlag; 1977.

10. Gazzaniga, M. S. *The Consciousness Instinct: Unraveling the Mystery of How the Brain Makes the Mind*. New York: Farrar, Straus and Giroux; 2018.

11. Baumeister, R. F. *The Self Explained. Why and How We Become Who We Are*. New York: The Guilford Press; 2022.

12. Kihlstrom, J. F., and Cantor, N. 'Mental Representations of the Self.' *Advances in Experimental Social Psychology*. 1984; 17(C): pp. 1–47.

13. Kuhn, M. H., and McPartland, T. S. 'An Empirical Investigation of Self-Attitudes.' *American Sociological Review*. 1954; 19(1): pp. 68–76.

14. McAdams, D. P. 'The psychology of life stories.' *Review of General Psychology*. 2001; 5: pp. 100–122.

15. Northoff, G., and Bermpohl, F. 'Cortical midline structures and the self.' *Trends in Cognitive Sciences*. 2004; 8(3): pp. 102–107.

16. Vogeley, K., and Gallagher, S. 'Self in the Brain.' In: *The Oxford Handbook of the Self*. Oxford: Oxford University Press; 2011: pp. 111–136.

17. Eustache, M. L., Laisney, M., Juskenaite, A., et al. 'Sense of

identity in advanced Alzheimer's dementia: a cognitive dissociation between sameness and selfhood?' *Consciousness and Cognition*. 2013; 22(4): pp. 1456–1467.

18. Klein, S. B. *The Two Selves. Their Metaphysical Commitments and Functional Independence*. Oxford: Oxford University Press; 2014.

19. Klein, S. B, and Gangi, C. E. 'The multiplicity of self: neuropsychological evidence and its implications for the self as a construct in psychological research.' *Annals of the New York Academy of Sciences*. 2010; 1191: pp. 1–15.

20. Minsky, M. *The Society of Mind*. New York: Simon & Schuster; 1987.

21. Tajfel, H., and Turner, J. C. 'An integrative theory of intergroup conflict.' In: Austin, W. G., and Worchel, S., eds. *The Social Psychology of Intergroup Relations*. Monterey, CA; 1979: pp. 33–37.

22. Jenkins, R. *Social Identity*. Abingdon: Routledge; 2014.

23. Tajfel, H., and Turner, J. C. 'The social identity theory of intergroup behavior.' In: Worchel, S., and Austin, W., eds. *Psychology of Intergroup Relations*. Chicago: Nelson Hall; 1986: pp. 7–24.

24. Carr, D. 'Personal identity is social identity.' *Phenomenology and the Cognitive Sciences*. 2021; 20(2): pp. 341–351.

25. Baumeister, R. F., and Leary, M. R. 'The need to belong: desire for interpersonal attachments as a fundamental human motivation.' *Psychological Bulletin*. 1995; 117(3): pp. 57–89.

26. Dunbar, R., Barrett, L., and Lycett, J. *Evolutionary Psychology. A Beginner's Guide*. London: Oneworld Publications; 2007.

27. Holt-Lunstad, J., Smith, T. B., and Layton, J. B. 'Social relationships and mortality risk: a meta-analytic review.' *Public Library of Science Medicine*. 2010; 7(7):e1000316.

28. Wang, F., Gao, Y., Han, Z., et al. 'A systematic review and meta-analysis of 90 cohort studies of social isolation, loneliness and mortality.' *Nature Human Behaviour*. 2023; 7(8).

29. Williams, K. D., and Zadro, L. 'Ostracism: On Being Ignored, Excluded, and Rejected.' In: Leary, M. R., ed. *Interpersonal Rejection*. Oxford: Oxford University Press; 2012: pp. 21–54.

30. International Organization for Migration. *World Migration Report 2022*. Geneva; 2022.

31. Hofstede, G. 'Dimensionalizing Cultures: The Hofstede Model in Context.' *Online Readings Psychology and Culture*. 2011; 2(1). https://doi.org/10.9707/2307-0919.1014

32. Winder, R. *Bloody Foreigners*. London: Little, Brown Book Group Limited; 2013.

Index